I0050274

QUANTITY SURVEY

Disclaimer (Exclusive clause)

The author and all employees disclaim any incorrect interpretation, wrong answers to questions or text, harm including emotional, psychological, and physical or any form of the liability to the reader/listener or being given information by third parties.

Furthermore, this disclaimer protects all contributory people, directors, employees, 3rd parties, authors and will not be liable for any injury caused.

Summary   Content

Module 1

Module 3

Dedication

I dedicate this book to myself. It reminds me of a business entrepreneur.

Acknowledgement.

As with all the previous one hundred and forty-eight books I have written and published, I acknowledge the Guardians' help, knowledge, wisdom, and the direction Almighty God has given me to draft this book. It would have been impossible to finish authoring these books because when I started writing,

I never knew how and when I would end or where to conduct my research, but once I began, I received more guidance from Almighty God. With him, this book was written. He directed and gave me the resources; therefore, the real author of this publication is God.

Forgive and love every then, trust, **believe and have faith in God. This FORMULA IS THE KEY TO HEAVEN.**

PREFACE

Quantity surveying is a critical discipline within the construction and infrastructure development industry. It involves managing all building and civil engineering project costs, ensuring value for money while achieving the required standards and quality. Quantity surveyors are pivotal in projects' financial and contractual management, from initial estimates to final accounts. This section explores the foundations of quantity surveying, its historical development, roles, skills required, and the modern tools used in the profession.

Historical Quantity surveying has its roots in ancient construction practices, where architects and builders estimated costs for structures like temples, castles, and monuments. The profession became formalised during the Industrial Revolution in the 19th century when the need for accurate cost management in large-scale construction projects grew. Modern quantity surveying emerged as a response to increasingly complex projects requiring specialised expertise in cost estimation, contract law, and financial management.

About the Author

My life experience—from being an abandoned child eating from bins/gutters in Ghana to being one of the most successful businessmen in Europe—motivates me to write this book.

I have visited most European countries and wish to share my experiences with my readers. I have also written over one hundred forty books and have many qualifications listed at the end of this book.

I have about six degrees, four post-graduates, including " institutional management, ten diplomas, 15 certificates, ect.

All are listed in this book.

Table of Contents

MODULE ONE

CHAPTER 1:

INTRODUCTION TO QUANTITY SURVEY

Quantity surveying is a critical discipline within the construction and infrastructure development industry. It involves managing all building and civil engineering project costs, ensuring value for money while achieving the required standards and quality. Quantity surveyors are pivotal in projects' financial and contractual management, from initial estimates to final accounts. This section explores the foundations of quantity surveying, its historical development, roles, skills required, and the modern tools used in the profession.

1. Historical Background of Quantity Surveying

Quantity surveying has its roots in ancient construction practices, where architects and builders estimated costs for structures like temples, castles, and monuments. The profession became formalised during the Industrial Revolution in the 19th century when the need for accurate cost management in large-scale construction projects grew. Modern quantity surveying emerged as a response to increasingly complex projects requiring specialised expertise in cost estimation, contract law, and financial management.

Data Required for the Preparation of an Estimate or Quantity Survey

Drawings

• Comprehensive and fully dimensioned architectural drawings, including plans, elevations, sections, and detailed layouts, are essential for accurately assessing quantities and measurements of materials and components involved in the construction project.

• These drawings serve as the foundation for visualising the project scope and calculating dimensions effectively.

Specifications

- Detailed written specifications outlining the type, quality, and standards of work to be carried out, along with descriptions of materials, their grades, proportions, and preparation methods, are required.

- These specifications ensure clarity and consistency in material selection, construction techniques, and compliance with engineering and safety standards.

Rates

- Updated cost rates for materials, labour (both skilled and unskilled), equipment usage, and transportation charges should be available to estimate the overall project cost accurately.

- The rates must reflect local market trends and include allowances for inflation or price variations, enabling precise budgeting and contract negotiations.

Measurements from Actual Work

- Quantities can also be derived from on-site measurements of completed work to validate estimates and manage progress payments.

- Different formulas are used for quantity calculations based on the type of work, including:

    o Volume Calculation: Quantity = Length × Width × Height (or Thickness).

    o Area Calculation: Quantity = Area of Cross-section × Length.

    o Surface Area Calculation: Quantity = length × width.

    o Linear Measurement: Quantity = length.

    o Unit Count: Quantity = Number of Units.

    o Weight Measurement: Quantity = Weight (for materials like steel or reinforcing bars).

Types of Estimates and Quantity Surveys

Preliminary or Approximate Estimate

    o This type of estimate provides a quick approximation of the cost of a proposed project. It is commonly used during the initial planning to help clients or decision-makers determine administrative approval and funding feasibility.

    o The estimate is usually based on comparisons with similar completed projects and employs area-based or volume-based calculations (e.g., cost per square meter or cubic meter).

o   Additionally, approximate quantities of materials and labour required per square meter can be calculated to give a rough idea of the needed resources.

Detailed Estimate

•   Once administrative approval is secured, a comprehensive and detailed estimate is prepared before inviting tenders for the work.

•   The project is broken down into smaller components or sub-works, and the quantities for each sub-work are calculated separately using dimensions taken from project drawings such as plans, sections, and elevations.

•   This estimate ensures accuracy and accountability, forming the basis for budget planning and procurement.

Quantity Estimates

•   This is a complete and itemised estimate of all the quantities required throughout the execution phase of the project.

- It provides precise details about the materials, labour, and equipment needed to complete each task, ensuring cost control and proper resource allocation during implementation.

Revised Estimate

- A revised estimate is prepared when the original estimate is exceeded by more than 5% due to unexpected circumstances, such as price fluctuations, design modifications, or errors in the initial estimate.

- It helps reassess the budget, ensuring transparency and enabling fund adjustments to keep the project on track.

Maintenance Estimate

- This estimate calculates the costs and quantities required for repairing, renovating, or maintaining existing structures such as roads, buildings, or bridges.

- It includes provisions for routine maintenance, emergency repairs, and periodic upgrades, ensuring the structure's longevity and functionality.

Supplementary Estimate

• A supplementary estimate is prepared when additional work not included in the original estimate arises during construction.

• It accounts for extra features or extensions needed and ensures that the project scope is expanded without delays in funding or approvals.

2. Roles and Responsibilities of a Quantity Surveyor

Quantity surveyors are involved in various stages of construction projects, ensuring that resources are used efficiently. Their responsibilities span from initial planning to project completion, providing oversight to control costs and mitigate risks. Key responsibilities include:

Cost Planning and Management;

Developing preliminary cost estimates based on project specifications and updating budgets as the project progresses. They track expenditures and highlight deviations, ensuring financial discipline and preventing cost overruns. Their ability to forecast costs accurately helps stakeholders make informed decisions.

Feasibility Studies;

Conducting comprehensive feasibility studies to assess whether proposed projects are financially and technically viable. This involves analysing site conditions, material requirements, labour costs, and regulatory factors. Their evaluations assist clients in determining whether to proceed with projects.

Tendering and Procurement:

Prepare detailed tender documents, including bills of quantities, that outline material specifications and labour requirements. Evaluate contractor bids based on cost, compliance, and quality, ensuring the selection of competent and cost-effective contractors.

Contract Administration;

Overseeing contract implementation to ensure compliance with legal and technical standards. They handle disputes, draft contract variations, and facilitate amendments when changes arise, maintaining project continuity and fairness.

Valuation and Payments;

Conducting valuations to assess the work completed and certifying interim payments to contractors. At project completion, they prepare final accounts to confirm the total expenditure and close financial records.

Risk Management

Identifying potential risks, such as inflation, delays, or material shortages. They devise strategies to mitigate these risks through contingency planning, insurance advice, and regular monitoring of market trends.

3. Skills and Competencies Required

To excel in quantity surveying, professionals must possess technical and interpersonal skills that enable them to navigate complex construction environments effectively.

Analytical Skills;

Evaluating large volumes of data to identify patterns and discrepancies helps surveyors make informed decisions about costs, contracts, and resources. Strong analytical abilities are essential for accurate forecasting and problem-solving.

Numerical Proficiency

Performing mathematical calculations and creating financial models with precision. Quantity surveyors handle complex budgets and cost estimations, requiring high levels of numerical accuracy.

Communication Skills;

Effectively communicating with architects, engineers, contractors, and clients. They must present technical data clearly, write detailed reports, and negotiate terms confidently.

Attention to Detail;

Ensuring the accuracy of cost estimates, contracts, and legal documentation to prevent disputes and losses. Errors in calculations or reports can lead to financial and legal challenges.

Knowledge of Construction Processes

Understanding building techniques, materials, and regulations to assess designs, recommend alternatives, and enforce compliance with quality standards.

4. Tools and Technologies in Quantity Surveying

Modern quantity surveying heavily relies on advanced tools and software to enhance accuracy and efficiency in cost management and project planning.

Building Information Modelling (BIM) integrates 3D modelling with cost and schedule data, enabling surveyors to visualise structures and make cost-related decisions in real time. It facilitates collaboration between stakeholders and reduces errors.

Cost Estimation Software;

Programs like Cost X and Win QS streamline budgeting, tender preparation, and reporting. These tools offer templates and automated calculations, saving time and improving precision.

Spreadsheet Applications;

Excel and similar programs are used for detailed calculations, data analysis, and financial tracking. Their versatility makes them indispensable for managing budgets and reports.

Project Management Software;

Tools such as Primavera and Microsoft Project help surveyors schedule activities, allocate resources, and monitor progress against timelines to ensure projects stay on track.

Important of quantity survey

a.   **Accurate Cost Estimation**

Quantity surveying is critical in estimating the probable construction cost before the project begins. It includes evaluating the costs of materials, transportation, labour, scaffolding, tools, equipment, water, taxes, and contractor profits.

Such estimates are essential for budgeting, project feasibility studies, and preparing tender documents to invite bids and secure contracts.

b.   **Material and Labour Planning**

A quantity survey provides precise calculations of the quantities of materials and labour requirements needed to complete a project successfully.

It enables effective procurement planning to avoid shortages or wastage and ensures the availability of resources at the right time, maintaining smooth workflow and productivity.

c.   **Monitoring and Evaluation of Work Progress**

Quantity surveying helps track work progress during and after construction. It ensures that the work aligns with the planned specifications and quality standards.

Contract payments are processed based on accurately completed task measurements, minimising disputes and ensuring fair compensation.

d.   **Legal and Financial Advisory Services**

A detailed quantity survey is valuable for providing clients with professional advice on:

o   Valuation of properties for sales, purchases, or mortgages.

o   Fixation of standard rent to ensure fair pricing for leased properties.

o   Insurance assessments and claims for damages in case of accidents or natural disasters.

o   Dispute resolution by offering independent valuations and expert assessments.

e.   **Risk Management and Cost Control**

- Quantity surveyors identify potential risks and uncertainties that could impact costs, such as price fluctuations, design changes, and unexpected delays.

- They assist in cost control throughout the project lifecycle by comparing actual expenditures with estimated budgets, enabling timely adjustments to avoid overspending.

f.   **Enhancing Efficiency and Sustainability**

- Quantity surveyors promote efficiency through detailed project planning and advanced technologies like Building Information Modelling (BIM) and cost estimation software.

- They also encourage the adoption of sustainable materials and eco-friendly practices to meet environmental standards while maintaining cost-effectiveness.

5. Challenges in Quantity Surveying

The profession is not without its challenges, including:

Cost Escalations

Managing unexpected increases in material prices, labour rates, and transportation costs. Surveyors develop contingency plans and negotiate contracts to account for fluctuations, mitigating financial risks.

Legal Disputes

Disagreements over contract terms, delays, and payments often lead to disputes. Surveyors draft clear, detailed contracts and employ dispute-resolution techniques to handle conflicts effectively.

Sustainability Concerns;

Balancing cost efficiency with environmentally friendly practices. Surveyors integrate renewable materials, recycle resources, and minimise waste to meet sustainability goals.

6. Future of Quantity Surveying.

The field is evolving rapidly with technological advancements and shifting industry priorities. Emerging trends include:

Sustainable Construction;

Emphasising green building practices and renewable materials to reduce environmental impact. Surveyors play a key role in evaluating sustainable options within budget constraints.

Artificial Intelligence;

Leveraging AI to predict costs, assess risks, and optimise resource allocation. AI enhances accuracy and speeds up decision-making.

Globalisation;

Managing international projects with diverse legal frameworks, cultural practices, and currency fluctuations. Quantity surveyors need to adapt to global standards and market conditions.

Contracts

A contract is a formal agreement between two or more parties that creates legally binding obligations. It requires each party to fulfil their agreed responsibilities.

Employer's Obligations:

1.   Appoint an engineer to manage the contract.

2.   Provide the site for the work.

3.   Supply necessary information, permits, and approvals.

4.   Ensure timely payment and provision of funds as per the contract.

5.   Participate in discussions with the engineer to resolve claims or disputes.

Contractor's Obligations:

1.   Carry out and complete the work, addressing any defects as needed.

The contractor must provide:

i.   Labour, materials, machinery, and equipment required.

ii.   Progress reports.

iii.   A work program and update it as necessary.

iv.   Set up the work area.

v.   Measure and/or assist the engineer in measuring.

vi.   Keep records of personnel and equipment.

vii.   Provide material samples as specified.

viii.   Conduct tests and re-tests.

ix.   Temporary works.

x.   j. Facilities for other contractors working on-site.

xi.   k. Maintain a clean site and remove waste.

2.   The contractor is required to:

a)   Sign the contract when requested.

b)   Obtain and submit securities, guarantees, and insurance policies.

c)   Ensure their representatives are available on-site at all times.

d)   Prepare and submit documents, including "as-built drawings" and operation/maintenance manuals.

e)   Follow the engineer's instructions.

f)   Allow the employer's personnel to access the site.

g)   Prepare and submit payment statements and documentation.

h)   Expose works for inspection when needed.

i)   Correct any defective work.

j)   Protect the employer from any claims or provide compensation.

k)   Notify the engineer of potential circumstances that may lead to claims.

l)   Obtain approval before assigning subcontractors or partners.

m)   Participate in consultations with the engineer.

n)   Comply with applicable laws, labour regulations, and other local laws.

Types of Contracts

Measured or Unit Rate Contract

In this type of contract, the cost is calculated by multiplying the quantities of work performed by the unit rates provided by the contractor in their bid. These rates are typically listed in the Bill of Quantities (BOQ). This type of contract is commonly used when significant changes in quantities or working conditions are expected. When reasonable quantity discrepancies arise, the contract can be paid based on the measured amounts multiplied by the agreed unit rates.

Advantages:

1. **Suitability:**

This contract is often used in large-scale projects funded by public bodies or governments. It is suitable for works that can be divided into separate items, with each item's quantity estimated accurately.

2. The employer only pays for the actual work completed.

3. The contractor typically accounts for a margin of variation and has a straightforward process for evaluating such changes.

4. The employer/engineer can provide additional drawings during the project execution.

Disadvantages:

1. The employer cannot fully predict the project's total cost until completion. The final price may vary significantly if the BOQ quantities are inaccurate or rough. Due to uncertainties in the quantities, the contractor may bid at unbalanced rates.

2.　　The engineer and contractor must perform detailed calculations and maintain records throughout the project.

3.　　Additional work or changes often lead to disputes, with the contractor possibly demanding higher rates than initially tendered.

Lump Sum Contract

In a lump sum contract, the contractor agrees to complete the project for a fixed amount as specified in the drawings and documents. Occasionally, provisions for adjusting the lump sum are made to accommodate additional work and minor changes. Usually, a BOQ is not provided, but if included, it serves only as a guide and not a contractual document. Instead, a schedule of rates may be used to assess the cost of any extra work.

Advantages:

1.　　From the employer's perspective, if no additional work is anticipated, the contract sum provides a clear and fixed cost for the project, which is particularly useful when operating within a strict budget.

2.   From the contractor's perspective, since they often prepare the design, they can increase their profit margin through effective planning and management and better control of the project timeline.

3.   Both parties require fewer resources for bookkeeping and accounting.

Disadvantages:

1.   Lump sum contracts require complete, detailed plans and specifications or the "Employer's Requirements" to be clear and comprehensive.

2.   Variations in the scope of work can lead to disputes about whether specific tasks are included in the original scope.

3.   This contract type may not be suitable for projects where the scope or nature cannot be accurately predicted, which could result in unfair outcomes for either party.

Cost-Plus Contract

A cost-plus contract differs from the measured and lump sum types as the employer agrees to pay the contractor for the actual costs incurred plus a fixed percentage of those costs to cover overhead and profit. The contractor is expected to execute the work per the drawings and specifications, with potential adjustments made as the project progresses.

Advantages:

1.   Work can begin before design and estimates are finalised, with quick decision-making and flexibility to accommodate changes that meet the employer's needs.

2.   The contractor is motivated to ensure high-quality work since their performance directly benefits the employer.

3.   No conflicts arise regarding extra work or omissions.

Disadvantages:

1.   The final cost to the employer is unpredictable.

2.   Both parties must manage extensive accounting for labour, materials, equipment, and other expenses.

3.   The contractor may lack the incentive to finish the work quickly or efficiently.

Suitability:

Despite some challenges, this contract is suitable for:

- Emergency projects require swift construction where no time is available for detailed drawings.

- Special or high-cost projects, such as palaces, where the focus is on material quality and craftsmanship rather than cost.

An alternative to the cost-plus contract is the cost-plus-fixed-fee contract, in which the contractor is paid the actual cost of construction plus a fixed fee for overhead and profit that does not fluctuate with the project's cost.

Construction Management Contract (C.M.)

In this contract, the employer hires a specialised construction manager (C.M.) to oversee and manage the project. The C.M. controls costs and timelines and manages the budget and schedule. The C.M. is typically paid on a staff-reimbursement basis. The C.M. helps select design consultants and contractors for the various packages (e.g., structural, finishes, electromechanical). The design remains with the professionals, but the C.M. plays a crucial role in control, coordination, certification, and resolving disputes.

CHAPTER 2:

CONSTRUCTION TECHNOLOGY AND MATERIALS

Construction technology refers to the techniques, methods, and tools used in building structures, infrastructure, and facilities. It involves the application of engineering principles, modern equipment, and innovative solutions to improve efficiency, safety, and sustainability.

Construction materials and technologies determine structures' quality, durability, and cost-effectiveness. Understanding these aspects is vital for quantity surveyors to evaluate costs, recommend materials, and plan projects effectively.

Classification of Building Materials

1. Natural Materials

- **Stone—Stone is used for foundations, walls, and paving due to its high compressive strength and resistance to weathering. It provides excellent durability, making it ideal for load-bearing structures, but** requires skilled labour for shaping and placement.

- **Wood**

Widely used in roofing, frames, and interior finishes. It is lightweight, flexible, and easy to shape, making it ideal for decorative purposes. However, it needs treatment to prevent termites, moisture, and fire damage.

- **Clay and Bricks**

Made from moulded clay and fired in kilns, these materials offer good thermal insulation, fire resistance, and affordability. They are versatile and used in walls, partitions, and flooring.

2. Manufactured Materials

- **Concrete**;

Concrete is the most commonly used material in construction. It comprises cement, sand, aggregates, and water and can be moulded into any shape. It provides excellent compressive strength, durability, and fire resistance. Pre-cast concrete elements, like slabs and beams, further improve efficiency.

- **Steel**;

Due to its tensile strength and flexibility, steel is a key material for reinforcement and structural frameworks. It is ideal for skyscrapers, bridges, and large spans but requires protective coatings to resist corrosion.

- **Glass**

Glass is often used in windows, facades, and skylights for aesthetic appeal and natural lighting. Modern advancements allow it to be shatter-resistant, tinted, or coated for insulation.

3. Composite Materials

- **Fibber-reinforced polymers (FRP) are lightweight** yet strong materials for reinforcing structures. They are corrosion-resistant and ideal for harsh environments.

- **Pre-stressed Concrete**

Incorporates steel cables or rods under tension, enhancing tensile strength. This technology is widely used in bridges and large-span structures, providing structural stability and load-bearing capacity.

Modern Construction Technologies.

Prefabrication and Modular Construction

- **Definition**

Components of a structure are manufactured offsite in controlled environments and transported to the site for assembly.

- **Advantages**

Reduces construction time, improves quality control, minimises waste, and lowers labour costs.

- **Applications**;

Prefabrication is used in residential housing, offices, hospitals, and temporary structures. It is instrumental in remote areas where on-site construction may be challenging.

3D Printing in Construction

- **Overview**;

Automated systems deposit materials layer by layer to create structures based on digital designs.

- **Benefits**

It enables fast and cost-efficient construction, reduces material waste, and allows for complex architectural designs that are difficult to achieve with traditional methods.

- **Limitations**

Requires specialised skills and investment in advanced equipment, limiting accessibility for small-scale projects.

Green Building Technology

- **Concept**

Focuses on sustainable construction practices that reduce environmental impact.

- **Examples**

Solar panels, rainwater harvesting systems, and recycled materials improve energy efficiency and resource conservation.

- **Benefits**

Lowers operational costs, promotes eco-friendliness, and improves occupant health and comfort.

Smart Construction Technologies.

- **Internet of Things (IoT)**

Sensors monitor structural health, detect faults, and enable preventive maintenance.

- **Building Information Modelling (BIM)**

allows the integration of design, scheduling, and cost data into a single digital model, streamlining planning, communication, and execution.

Properties of Construction Materials.

1. **Physical Properties**
- **Density**

Measures mass per unit volume, affecting weight and strength. Dense materials like concrete provide durability, while lighter materials like wood are easier to handle but may require reinforcement.

- **Porosity**

determines water absorption capacity, which impacts durability. Materials with high porosity, such as bricks, may need waterproofing treatments in wet environments.

- **Thermal Conductivity**;

It influences insulation capabilities, with wood and foam providing better thermal insulation than metals or concrete.

2. **Mechanical Properties**
- **Strength**

Measures the ability to resist forces without deformation. High-strength materials like steel and concrete are essential for load-bearing structures.

- **Elasticity**;

Refers to returning to the original shape after removing stress. Elastic materials are helpful in areas prone to earthquakes or vibrations.

- **Durability**;

Indicates resistance to wear, corrosion, and weathering, ensuring longevity and reducing maintenance costs.

3.   **Chemical Properties**

•   **Corrosion Resistance**

Determines suitability for environments with moisture or chemicals. Materials like stainless steel and treated wood resist corrosion.

•   **Fire Resistance**;

This is essential for safety, especially in high-risk areas. Concrete and gypsum boards provide excellent fire protection.

Quality Control and Testing in Materials

1.   **Material Testing Methods.**

•   **Destructive Testing**

Samples are broken or stressed to determine strength, durability, and load capacity. Compression tests for concrete and tensile tests for steel are common examples.

•   **Non-Destructive Testing (NDT) -**
Techniques like ultrasonic testing and radiography detect internal flaws without damaging materials.

2.   **On-Site Testing**

- **Slump Test**;

Measures the workability of concrete, ensuring consistency and water-cement ratio.

- **Brick Absorption Test**

Evaluates water absorption capacity to ensure durability under moisture exposure.

3. **Compliance Standards**

- **ISO Standards**

Provide international quality, durability, and safety guidelines in materials and construction methods.

- **Local Building Codes**;

Specify acceptable practices based on environmental and geographic conditions, ensuring safety and compliance.

CHAPTER 3:

QUANTITY SURVEY ITEMS AND METHODS

Introduction

Quantity surveying involves determining the estimated materials needed for a project, typically by a professional surveyor or engineer. These estimated quantities are shared with potential bidders to enable them to submit their prices. In this bidding process, all contractors work with the exact amounts estimated. Contractors' estimators invest time in calculating unit prices for various items within the project. To secure the bid, contractors aim to minimise purchasing and installing materials costs.

As the project progresses, the actual quantities are compared with the amounts estimated. For instance, if the estimated concrete quantity for a wall is 23 m³, but the exact amount used is 26 m³, the contractor will be paid for the extra 3 m³.

If there's a significant discrepancy between the estimated and actual quantities, adjustments to the unit price may be necessary. Slight differences are usually adjusted at the contractor's original unit price, while more significant discrepancies may require renegotiation of the unit price.

If the contractor anticipates potential changes in the required quantities, they may adjust their bid accordingly. For example, if the contractor knows that the backfill material will change from excavated soil to base-course, they might quote a lower unit price for soil (e.g., 5 JD/m³) and a higher price for base-course (e.g., 15 JD/m³). If the assumed quantities were 2,000 m³ of soil and 100 m³ of base course, the bid would total 11,500 JD. However, if the amounts change to 100 m³ of soil and 2,000 m³ of base course, the contractor's revised price would be 30,500 JD, resulting in a higher profit due to the quantity alteration.

Bill of Quantities BOQ

A **Bill of Quantities (BOQ)**

is a detailed document in construction projects that provides a comprehensive list of the required materials, labour, and other resources. It outlines the quantities and unit prices of each work item to assist contractors in preparing their bids. The BOQ is a critical tool for the contractor and the employer to ensure clarity, transparency, and consistency in pricing and work execution.

Key Features of BOQ:

1.   Description of Work:

The BOQ clearly describes each work item involved in the project, specifying the type of materials, labour, and equipment required for each task.

2.   Quantities:

It provides the required quantities of materials and work, typically in meters, cubic meters, square meters, etc.

3.   Unit Prices:

For each item, the BOQ lists the unit price that a contractor may offer based on market rates and project requirements.

4.   Total Cost Calculation:

The BOQ estimates the total project cost by multiplying the number of work items by the unit prices. This helps the employer evaluate bids and allocate budgets.

5.   Standard Specifications:

The BOQ usually refers to specific construction standards and regulations, such as national or international codes and local government specifications, to ensure the work meets the required quality standards.

6.   Adjustments:

The BOQ may include provisions for adjustments in case of changes in the project scope or quantities. Contractors and employers can agree on adjustments to unit rates or amounts as necessary.

Purpose of BOQ:

- **Bidding**:

The BOQ enables contractors to submit competitive bids based on clear, quantified work descriptions and unit rates.

- **Cost Estimation**:

It is a tool for estimating the project's total cost, offering a transparent basis for comparing different contractors.

- **Project Control**:

- During the project execution, the BOQ acts as a reference to monitor progress, manage resources, and ensure that payments align with the work completed.

- **Quality Assurance**:

Using standard specifications in the BOQ ensures that all work is carried out according to the established norms and quality standards.

Advantages:

Provides precise and accurate cost estimation.

It helps to standardise the bidding process.

Ensures transparency and fairness in contract pricing.

Acts as a reference for monitoring work progress and quality.

Disadvantages:

- Requires accurate measurements and detailed specifications, which may delay initial stages.

- Changes in quantities or specifications during the project may require revisions to the BOQ.

A sample BOQ

Bill of Quantities (BOQ)

Project Name:

Residential Building

Project Address:

XYZ, City, Country

Client: ABC Construction Ltd.

Contractor: XYZ Builders Ltd.

Date: January 2025

Section 1: Civil Works

Item No.	Description of Work	Unit	Quantity	Unit Rate (Currency)	Total Cost (Currency)
.1	Excavation for foundation	3	150	10	1,500
.2	Concrete for foundation (Grade 25)	3	50	80	4,000
.3	Brick work (1st floor)	2	100	45	4,500
.4	Concrete slab for the first floor	2	120	50	6,000

Item No.	Description of Work	Unit	Quantity	Unit Rate (Currency)	Total Cost (Currency)
.5	Back filling with excavated soil	3	100	20	2,000

Section 2: Architectural Works

Item No.	Description of Work	Unit	Quantity	Unit Rate (Currency)	Total Cost (Currency)
.1	Brick wall construction (internal)	2	200	35	7,000

Item No.	Description of Work	Unit	Quantity	Unit Rate (Currency)	Total Cost (Currency)
.2	Plastering (walls)	2	500	15	7,500
.3	Painting (walls and ceiling)	2	600	8	4,800
.4	Floor tiling (ceramic)	2	100	25	2,500

Section 3: Mechanical Works

Item No.	Description of Work	Unit	Quantity	Unit Rate (Currency)	Total Cost (Currency)
.1	Air conditioning system installation	nit	5	1,500	7,500
.2	Plumbing works (pipes and fittings)		200	12	2,400
.3	Electrical wiring and installation		300	10	3,000

Section 4: Electrical Works

Item No.	Description of Work	Unit	Quantity	Unit Rate (Currency)	Total Cost (Currency)
.1	Wiring for lights and power outlets		500	,500	2
.2	Installation of switches and sockets	nit	50	50	1
.3	Distribution board installation	nit	1	00	8 00

Summary of Total Costs

Quantity Survey   J. Safo

Section	Total Cost (Currency)
Civil Works	20,000
Architectural Works	21,800
Mechanical Works	12,300
Electrical Works	3,450

Grand Total for the Project: 57,550 (Currency)

Method of Estimating Quantities

Several methods are used to determine the quantities of materials required in construction. The method chosen depends on the project type, the development stage, and the available information.

1. Direct Measurement Method:

- The most accurate method is measuring the quantities directly from the drawings.

- It is typically used for projects that have detailed designs and accurate measurements available.

2. Elemental Method:

- A method where the project is broken down into elements (e.g., foundation, superstructure, roofing).

- The quantities for each element are then estimated based on the general nature of the work.

3. Approximate Quantities:

- A rough estimate of quantities based on previous experience or standard practices.

- This method is used when detailed drawings are unavailable, or the project is in the early stages.

Adjustments and Variations

As construction progresses, actual quantities may differ from the estimates in the BOQ. Variations in work, material changes, or unforeseen circumstances may lead to adjustments.

Handling Variations:

- **Slight Variations**:

Minor changes in quantities or work are generally accounted for at the same unit rate initially bid by the contractor.

- **Large Variations**:

Significant changes in scope may require renegotiation of the unit rates or a new price agreement.

In cases where the contractor is aware of potential changes or deviations from the original quantities, they may adjust their bid to account for the uncertainty (e.g., adjusting unit prices for potential material changes).

Advantages and Disadvantages of Quantity Surveying Methods

Advantages:

- **Accurate Cost Estimation**:

This helps precisely estimate the project's total cost.

- **Cost Control**:

This helps track actual costs against estimated costs throughout the project, which helps prevent cost overruns.

- **Clear Communication**:

The BOQ clearly outlines the work scope, reducing ambiguity between contractors and clients.

Disadvantages:

- **Time-Consuming**:

Preparing a BOQ can be time-consuming, especially for large projects.

- **Changes in Scope**:

Variations during the project can complicate the initial estimates and lead to disputes.

- **Requires Expertise**:

The BOQ's accuracy depends on the quantity surveyor's expertise and the availability of detailed plans and specifications.

CHAPTER 4:

MEASUREMENT OF BUILDING WORKS

Measuring building works is essential to the construction process, ensuring accurate quantities are calculated for materials, labour, and time. These measurements serve multiple purposes, including cost estimation, billing, progress monitoring, and compliance with contract terms. Accurate measurement of building works provides a reliable basis for creating the **Bill of Quantities (BOQ)**, which outlines the cost breakdown for each aspect of the construction process.

Purpose of Measuring Building Works

- **Cost Estimation**: Accurate measurements are critical in the early stages of a project when preparing cost estimates. They help quantify the materials and labour needed for each task, facilitating the preparation of an accurate budget.

  ○ *Example*: Measuring a concrete slab's volume (length x width x thickness) allows for estimating the amount of cement, sand, gravel, and labour required.

- **Contractual Documentation**:

Measurement is a legally binding document defining the work to be done and its value. The measurements from construction drawings form the basis for agreements between clients, contractors, and subcontractors.

 o *Example*: A contractor may submit a bid based on an accurate take-off from the drawings and the BOQ, outlining unit prices for each item, such as excavation or brickwork.

- **Progress Valuation**:

Work is often paid for during construction based on the project's progress. Measurement helps track completed work, ensuring contractors are paid fairly for their progress.

 o *Example*:

In a large construction project, the contractor may be paid based on the percentage of the building completed, such as 60% of the brickwork or 70% of the roofing.

- **Variation Measurement**:

Changes in the original design require measuring additional or modified works. Variations in design or unforeseen site conditions lead to new measurements and adjustments in cost.

- ○ *Example*:

If unexpected structural support is needed during construction, the additional material and labour cost for this new work will be measured and added to the original contract value.

Key Principles of Measurement

- • **Accuracy**:

The precision of the measurement is essential to avoid discrepancies. Minor errors can lead to significant cost differences and affect project timelines.

- ○ *Example*:

If the length of a wall is mismeasured, the quantity of bricks needed could be overestimated, leading to the purchase of excess material.

- **Consistency**:

Using the same measurement approach throughout the project helps ensure uniformity, particularly when multiple contractors or teams are involved.

  - *Example*:

A contractor should apply the same method of calculating wall areas (length x height) across different building sections.

- **Standardisation**:

Following industry-standard measurement codes (e.g., RICS, SMM7) ensures that measurements are uniform and accepted across projects, reducing the likelihood of disputes.

  - *Example*:

The **New Rules of Measurement (NRM)** published by RICS outlines specific guidelines for measuring building works in the UK.

- **Transparency**:

Clear and understandable measurement methods allow all stakeholders (client, contractor, surveyor) to be on the same page, reducing the potential for misunderstandings.

  - *Example*:

If a surveyor notes that measurements were taken from the foundation plan rather than the architectural plan, it should be documented to ensure transparency.

Methods of Measurement for Building Works

Depending on the complexity of the building works, different methods are employed. These methods range from traditional manual measurement to modern digital tools and software.

Traditional Methods of Measurement

Traditional methods have been used for decades and involve manual calculations based on detailed drawings and specifications.

- **Direct Measurement**:

This method involves taking physical measurements directly from construction drawings or on-site. For example, a structure's length, width, and height (e.g., a wall or a floor slab) can be measured using a scale ruler or measuring tape.

- o *Example*:

73

If measuring a concrete slab, you may multiply the length and width from the drawings to get the total area (e.g., 10 meters x 20 meters = 200 square meters).

- **Estimation and Take-off**:

It involves "taking off" quantities from the construction drawings (e.g., walls, foundations, slabs, windows). This method estimates the material and labour needed for a particular task.

- *Example*: To estimate the amount of concrete needed for a foundation, the surveyor measures the length, width, and depth of the foundation trench and multiplies these dimensions to calculate the total cubic volume of concrete required.

Computer-Aided Measurement

Technology advancements have introduced new tools for measuring building works with greater accuracy and efficiency.

- **Building Information Modelling (BIM)**:

BIM integrates design and construction data into a three-dimensional model, allowing surveyors to extract quantities directly from the model. It enables real-time collaboration among all stakeholders and reduces human error.

- Example:

In BIM, a 3D model of a building can be developed, and the quantity surveyor can extract the total volume of concrete needed for foundations, the number of doors and windows, or the area of walls directly from the digital model.

- **Example Project**:

A contractor using BIM software could calculate the total number of bricks needed for a wall by automatically quantifying the wall's area from the 3D model and cross-referencing it with the size of standard bricks.

- **Computerised Quantity Surveying Software**:

  - Programs like **CostX** and **Bluebeam Revu** allow surveyors to measure quantities directly from digital drawings, which speeds up the process and increases accuracy.

  - Example:

With **CostX**, a surveyor can click on the digital drawing's walls, floors, and roof areas to automatically calculate quantities for concrete, timber, and other materials.

Common Building Works and Measurement Techniques

Building works are measured according to their type. Below are the standard building elements and their measurement methods:

Site Preparation and Earthworks

- **Excavation:**

Measured by volume (cubic meters).

  o *Example:*

Excavation for a foundation involves removing soil from a trench. If the trench is 10 meters long, 2 meters wide, and 1 meter deep, the volume of earth to be excavated is 10 x 2 x 1 = 20 cubic meters.

- **Grading:**

Measured by area and depth.

  o *Example:*

Grading involves levelling or sloping the site. If a building requires 200 square meters of grading at a depth of 0.2 meters, the volume of grading material is 200 x 0.2 = 40 cubic meters.

Concrete Works

- **Foundation Concrete**:

Measured by volume (cubic meters).

- *Example*:

A slab foundation measuring 10 meters in length, 5 meters in width, and 0.2 meters in depth has a volume of 10 x 5 x 0.2 = 10 cubic meters of concrete.

Brickwork and Blockwork

- **Walls**:

Measured by area (square meters).

- *Example*:

A brick wall measuring 5 meters in length and 3 meters in height would have an area of 5 x 3 = 15 square meters.

o   For cavity walls, the thickness of the wall (e.g., 0.3 meters) must also be taken into account, and calculations for internal and external surfaces are made accordingly.

o

Roofing

- **Roof Area**:

Measured by area (square meters).

o   *Example*:

For a pitched roof, if the building's footprint is 100 square meters and the roof has a 30-degree slope, the actual roof area will be greater than the footprint area. The location is determined by factoring in the hill.

- **Roof Trusses and Framing**:

Measured by quantity and weight.

o   *Example*:

The number of trusses needed for a gable roof is measured by counting the individual trusses, and the weight of each truss is calculated based on the material used (wood or steel).

Finishes and Decorations

- ## Painting and Coatings:

Measured by area (square meters).

- ○ *Example*:

If 100 square meters of wall surface requires painting and two coats of paint, the total area for painting will be 100 x 2 = 200 square meters of paint to be applied.

- **Tiling**: Measured by area (square meters).

- ○ *Example*: For a bathroom floor that is 5 meters by 4 meters, the area for tiling is 5 x 4 = 20 square meters.

- ○

Doors and Windows

- **Doors and Windows**: Measured by quantity or area (square meters).

- ○ *Example*:

A door measuring 2 meters in height and 1 meter in width has an area of 2 x 1 = 2 square meters. Similarly, a window of 1.5 meters by 1 meter would have an area of 1.5 square meters.

Necessary Measurement Standards and Codes

- **Royal Institution of Chartered Surveyors (RICS)**: RICS provides comprehensive standards for measuring building works, ensuring uniformity in the industry.

  o *Example*:

NRM 1 outlines the process of measuring building works for tendering purposes, ensuring contractors can provide accurate unit prices.

- **Standard Method of Measurement (SMM7)**:

The SMM7 is a well-established method in the UK for measuring building works and forming the basis for BOQs.

  o *Example*:

Under SMM7, measurements for concrete foundations, brick walls, and roofing elements are clearly defined, specifying how each item should be quantified.

Challenges in the Measurement of Building Works

- **Changes in Design**:

Alterations to design during construction require new measurements, impacting the original quantities and costs.

- Example:

If a change in the design results in an additional floor, the quantities of materials, such as concrete and steel, need to be measured and priced again.

- **Measurement Errors**: Manual measurement errors or misinterpretation of drawings can lead to discrepancies that affect material procurement or costs.

- Example:

If a surveyor misreads the height of a wall by even a few centimetres, the overall material quantities for brickwork may be inaccurate.

- **Unforeseen Site Conditions**:

Site conditions, such as soil instability or unexpected obstructions, can cause variations in quantities, requiring additional measurements.

- *Example*:

If soil conditions require a deeper foundation than initially planned, the additional excavation volume must be measured and accounted for in the cost estimation.

CHAPTER 5:

CONSTRUCTION ECONOMICS

Construction Economics

Construction economics refers to studying and applying economic principles and techniques to the construction industry. It deals with the financial aspects of construction projects, including cost management, resource allocation, pricing strategies, and the economic viability of projects. Construction economics is crucial to a project's success, guiding contractors, project managers, clients, and stakeholders to make informed financial decisions throughout the project lifecycle.

Key Areas of Construction Economics

Construction economics encompasses several core areas, each contributing to the overall financial success of construction projects.

Cost Estimation and Control

- **Cost Estimation**:

Predicting the costs associated with a construction project. It includes direct costs (e.g., labour, materials, equipment) and indirect costs (e.g., overhead, administration, contingency).

  o   *Example*:

Estimating the cost of building a residential house requires calculating the cost of materials such as bricks, cement, and steel, as well as labour costs for construction workers and the cost of machinery.

  •   **Cost Control**:

Once a project is underway, controlling costs becomes crucial to ensure that the actual expenses do not exceed the budgeted amounts.

  o   *Example*:

Implementing a robust cost control system can involve tracking expenses for each construction phase (e.g., foundation, framing, roofing) and comparing them to the initial estimates to ensure the project stays on budget.

Financial Feasibility

Financial feasibility studies assess whether a construction project is economically viable.

  o   *Example*:

A developer may conduct a financial feasibility study to determine if the projected rental income over a set period justifies the cost of constructing a shopping mall.

- Key factors in feasibility analysis:

  o **Capital investment:**

The initial cost required to complete the construction project.

  o **Operational costs:**

Ongoing costs, including maintenance, utilities, and staffing.

  o **Return on investment (ROI):**

The profit or income expected relative to the investment.

Procurement Strategies

- **Procurement** is acquiring goods, services, and construction works necessary for a project.

  o Standard procurement methods include:

  ▪ **Traditional Procurement:** The client hires an architect to design the project, and then a contractor is hired through tendering to build it. The contractor is at higher risk as they are paid for completed work.

- **Design and Build**:

A single contractor is responsible for both the design and construction, streamlining the process but potentially reducing the client's control over the design.

Construction Management at Risk:

The contractor manages the entire construction process and assumes the risk of delivering the project on time and within budget.

- **Public-Private Partnerships (PPP)**: A partnership between the government and a private entity, where the private party finances, builds, and operates a facility for a set period before transferring ownership to the government.

Cost-Benefit Analysis (CBA)

- **Cost-benefit analysis evaluates a construction project's economic value by comparing its** total costs to the expected benefits.

  - *Example*:

The costs of a road construction project would include the price of materials, labour, and equipment, while the benefits might consist of reduced travel time, lower fuel consumption, and increased economic activity in the region.

- The **Net Present Value (NPV)**

  method is often used in CBA to account for the time value of money. This method discounts future cash flows to their present value.

  Formula:

$$NPV = \sum \left(\frac{C_t}{(1+r)^t} \right)$$

Where:

- C_t is the cash inflow at time t,

- r is the discount rate,

- t is the period.

Project Financing

- **Project Financing** involves securing funds for the construction of a project. It can include loans, equity financing, or a combination of both.

   o   *Example*:

A construction company may take out a loan or seek investors to fund the development of a new high-rise building, with repayments made using the property's future rental income.

- Standard financing methods include:

   o   **Equity Financing**:

Raising capital by selling shares in the company or project.

   o   **Debt Financing**: Borrowing money from banks or other financial institutions to fund the project.

   o   **Joint Venture Financing**: Two or more entities combine resources to finance a project, sharing the risks and rewards.

Construction Market and Economic Cycles

The construction industry is significantly affected by the broader economic environment.

Economic cycles, interest rates, inflation, and government policies are critical in determining construction project demand and supply.

Economic Cycles

- **Boom Cycle**:

During an economic boom, demand for construction services increases. It leads to more projects and higher prices for materials and labour. Contractors may benefit from increased profits, but costs also rise.

  o *Example*:

During periods of rapid urbanisation, the demand for residential, commercial, and infrastructure development soars, leading to a boom in the construction market.

- **Recession Cycle**:

During economic downturns, construction activity slows. Government spending on infrastructure projects may decrease, and private investments in construction may become more cautious. It can lead to project delays, lower demand for labour, and cost-cutting measures by construction companies.

  o *Example*:

In the wake of the 2008 financial crisis, many countries experienced a slowdown in construction, especially in the housing and commercial sectors.

Interest Rates

- Interest rates are a key determinant of the cost of borrowing for construction projects. When interest rates are high, the cost of financing increases and construction firms and clients may delay or cancel projects.

  o *Example*:

If the central bank raises interest rates, a construction firm may face higher loan repayment costs, potentially leading to higher project costs or a reduction in the number of new projects initiated.

Inflation and Material Costs

Inflation affects the prices of raw materials, labour, and energy, which impacts the overall cost of construction projects.

  o *Example*:

A sudden increase in the price of steel or cement due to inflation can lead to cost overruns in a project with a fixed budget.

- **Material Supply Chain**:

   Global supply chain disruptions (e.g., caused by trade tariffs, pandemics, or natural disasters) can lead to price volatility in construction materials.

   o   *Example*: The COVID-19 pandemic led to global supply chain disruptions, significantly increasing the cost of materials like lumber, steel, and copper.

Economic Factors Influencing the Construction Industry

Several macroeconomic and microeconomic factors influence construction economics.

Understanding these is crucial for effective decision-making.

Government Policies and Regulations

- Government policies,

  including taxation, environmental regulations, and building codes, significantly shape construction economics.

  - *Example*:

A government introducing a carbon tax on building materials like concrete and steel can raise costs for construction companies, making sustainable alternatives more attractive.

- Regulations governing labour, safety, and environmental protection also influence the cost and feasibility of projects.

  - *Example*:

Stricter safety regulations or labour laws (e.g., minimum wage increases or limits on working hours) can increase labour costs and extend project timelines.

Technological Advancements

- Technological innovations such as Building Information Modelling (BIM), modular construction, 3D printing, and automation can reduce construction costs and improve productivity.

  o *Example*:

  BIM allows for better coordination of construction teams and minimises costly errors during construction, reducing overall costs.

- Prefabrication and modular construction methods enable faster project completion and reduced labour costs, which are especially beneficial in large-scale construction projects.

  o *Example*:

Modular homes, where parts of the structure are pre-built offsite and then assembled onsite, can save time and labour costs compared to traditional construction methods.

Challenges in Construction Economics

Despite its importance, construction economics faces several challenges that can impact the successful execution of a project.

Cost Overruns and Delays

- One of the most common challenges in construction economics is the occurrence of cost overruns and delays. Factors contributing to this include inaccurate cost estimation, changes in project scope, unforeseen site conditions, and poor project management.

  o *Example*:

A building project might face cost overruns if additional foundation work is required due to unexpected soil conditions not accounted for in the initial cost estimate.

Resource Shortages

- Labour, materials, or equipment shortages can disrupt the construction schedule and increase costs.

  o *Example*:

A shortage of skilled labour, such as electricians or carpenters, can delay a project and result in higher wages for available workers, increasing the overall cost.

Environmental and Sustainability Concerns

- Increasing pressure to adopt sustainable construction practices, such as energy-efficient buildings, renewable materials, and waste reduction, can increase project costs.

  o *Example*:

Using eco-friendly materials like recycled steel or sustainable timber can increase material costs. However, the long-term savings and benefits in terms of energy efficiency may offset this initial investment.

CHAPTER 6:

MATHEMATICS AND STATISTICS FOR SURVEYORS

Mathematics is critical in surveying, particularly in determining distances, angles, and areas. Surveyors frequently use triangles, azimuths, angles, and bearings to measure land, construction sites, and other locations.

The Right Triangle

A right triangle is a triangle with an angle of precisely 90 degrees. This type of triangle is fundamental to many surveying applications, particularly in calculating distances, elevations, and angles.

Key Concepts and Formulas

- **Pythagorean Theorem:**

This theorem is crucial for surveying when working with right triangles. It states that in a right triangle, the square of the length of the hypotenuse (the side opposite the right angle) is equal to the sum of the squares of the other two sides.

- ○ **Formula:**

$$a^2+b^2=c^2$$

Where:

- a and b are the lengths of the two legs (sides that form the right angle),

- c is the length of the hypotenuse (the side opposite the right angle).

Application in Surveying:

- **Horizontal Distances**:

When surveyors need to calculate the horizontal distance between two points on uneven terrain, they can use the Pythagorean Theorem. The horizontal distance can be calculated if the vertical distance (elevation difference) and the slant distance are known.

○   *Example*:

If a surveyor knows the vertical distance between two points is 20 meters and the hypotenuse is 25 meters, the horizontal distance can be calculated using:

$$a^2 = c^2 - b^2 \implies a^2 = 25^2 - 20^2 \implies a = \sqrt{625 - 400} = \sqrt{225} = 15\,\mathrm{m}$$

In this example, the horizontal distance is 15 meters.

Trigonometric Ratios:

$$\text{Sine (sin)} = \frac{\text{opposite}}{\text{hypotenuse}}$$

$$\text{Cosine (cos)} = \frac{\text{adjacent}}{\text{hypotenuse}}$$

$$\text{Tangent (tan)} = \frac{\text{opposite}}{\text{adjacent}}$$

These ratios are helpful in surveying when determining unknown distances or angles when sure sides of a right triangle are known.

Oblique Triangles

An oblique triangle is any triangle that does not contain a right angle. These triangles can be **acute** (all angles less than 90°) or **obtuse** (one angle greater than 90°). Surveyors frequently encounter oblique triangles in fieldwork, especially when measuring distances and angles that are not right-angled.

Key Concepts and Formulas

- **Law of Sines**:

   This law is beneficial when surveying an oblique triangle, where two angles and one side are known (or two sides and an angle are known).

   o **Formula**:

$$\frac{\sin A}{a} = \frac{\sin B}{b} = \frac{\sin C}{c}$$

where:

- $A, B,$ and C are the angles of the triangle,

- $a, b,$ and c are the sides opposite these angles.

Law of Cosines:

This law helps calculate unknown sides or angles in an oblique triangle when two sides and the included angle are known or when all three sides are known.

Formula:

$$c^2 = a^2 + b^2 - 2ab \cdot \cos C$$

where:

- a, b, and c are the sides of the triangle,
- C is the included angle.

Application in Surveying:

- ### Triangulation:

Surveyors often use oblique triangles to measure the angles and sides of a triangle to determine the positions of unknown points. For example, if a surveyor knows the distances between points A and B and the angle between them, they can use the Law of Sines or the Law of Cosines to find the unknown distances or angles.

- ○ *Example*:

Given two known points and the angle between them, surveyors can use the Law of Sines to calculate the location of a third point.

Azimuths, Angles, & Bearings

Azimuths, angles, and bearings are essential in determining directions and positions in surveying.

These concepts represent the orientation of lines or points relative to a reference direction, typically true north or magnetic north.

Key Concepts

- **Azimuth**:

An azimuth is the angle measured from a fixed point (usually north) in a clockwise direction. It represents the direction of one point relative to another. Azimuths are typically measured in degrees from 0° to 360°.

  ○ **Example**:

 A surveyor measures an azimuth of 90° to represent the direction due east. If the azimuth is 180°, the direction is due south.

- **Bearings**:

Bearings represent directions relative to either the north or south axis. Bearings are often expressed in angles from a specific direction (north or south) to the east or west. Bearings are written as angles, e.g., N 45° E, meaning 45 degrees east of north.

  ○ **Example**:

A bearing of S 30° E means a line goes 30 degrees to the east of the south direction.

  ○ **Quadrants**:

Bearings are usually divided into four quadrants:

- **Northeast (NE)**: 0° to 90°

- **Southeast (SE)**: 90° to 180°

- **Southwest (SW)**: 180° to 270°

- **Northwest (NW)**: 270° to 360°

- **Horizontal Angles**:

  o **Horizontal angles** define the direction between two points about a baseline or reference direction. These are critical in establishing the layout of land or a construction project.

  o *Example*:

Surveyors measure horizontal angles using instruments like the theodolite to determine the direction from one point to another.

Application in Surveying:

- **Determining Locations and Directions**:

Surveyors use azimuths and bearings to map out land accurately, set boundaries, and navigate construction sites. Surveyors can triangulate positions and establish the orientation of new construction projects by using known points and measuring azimuths or bearings.

- **Measuring Angles for Boundary Lines**:

In land surveying, the surveyor may establish or verify boundary lines for a piece of property. By measuring the azimuths or bearings of each side, the surveyor can plot the exact location of property lines, ensuring they are accurate and legally binding.

Coordinate Geometry (COGO)

Coordinate Geometry (COGO) is a branch of geometry used extensively in land surveying, mapping, and engineering to solve geometric problems using coordinates. It involves determining the positions of points and distances between them, along with other geometric properties such as slopes, angles, and areas, using the principles of Cartesian geometry.

Key Concepts and Applications:

- **Cartesian Coordinate System**:

Points are identified by pairs of coordinates (x, y) on a 2D plane or (x, y, z) in 3D space.

- **Distance Between Two Points**:

- The distance d between two points P (x1, y1) P (x_1, y_1) P (x_1, y_1) and Q (x_2, y_2) Q (x_2, y_2) Q (x_2, y_2) is given by the formula:

$$d = \sqrt{(x_2 - x_1)^2 + (y_2 - y_1)^2}$$

Angle Between Two Points:

The angle θ between two points concerning the x-axis can be calculated using the formula:

$$\theta = \tan^{-1}\left(\frac{y_2 - y_1}{x_2 - x_1}\right)$$

Applications in Surveying: COGO is applied to set boundaries, define alignments in roads or railways, and establish the locations of various survey markers on a land parcel.

Law of Sines

The **Law of Sines** is used in trigonometry to solve for unknown sides or angles in non-right-angled triangles (oblique triangles). It is beneficial in surveying when working with triangles where you know either two angles and one side or two sides and one angle.

Formula:

$$\frac{\sin A}{a} = \frac{\sin B}{b} = \frac{\sin C}{c}$$

Where:

- A, B, and C are the angles of the triangle,
- a, b, and c are the sides opposite to these angles.

Applications in Surveying:

- ### Triangulation:

Surveyors use the Law of Sines to calculate the distance between two points, even if the distance is not directly measurable.

- ### Example:

If you know two angles and one side of a triangle, you can use the Law of Sines to find unknown distances or angles in land surveys.

Bearing, Bearing Intersections

Bearings describe the direction of one point relative to another and are essential in establishing the positions and orientations of boundary lines.

Key Concepts:

- **Bearing**:

A bearing is typically given as an angle measured from the north or south toward the east or west. Bearings are usually expressed in degrees, with 0° being due north, 90° due east, 180° due south, and 270° owing west.

- **Bearing Intersection**:

Bearing intersections are used to locate the point where two lines intersect. By knowing the bearings of two lines and the distance between points along those lines, surveyors can use geometry to calculate the intersection point.

Applications in Surveying:

- **Land Surveying**:

Surveyors use bearings to establish boundary lines and points. By measuring the bearings from two known points to a third, they can determine the exact intersection point, essential for boundary determination.

Bearing, Distance Intersections

When both **bearings** and **distances** are known, surveyors can calculate the intersection point of two lines. This process involves solving for unknown points based on known distances and directions.

Key Concept:

- **Intersection Formula**:

- You can calculate the intersection point if you know the bearings and distances between two points (say AAA and BBB) and the distance from these points to a third point, CCC.

  o The principle involves applying the law of sines and cosines with geometric calculations.

Applications in Surveying:

- **Property Boundaries**:

Surveyors calculate where property boundary lines intersect using known bearings and distances from markers. These intersections define property corners or other vital locations on a parcel of land.

Law of Cosines

The Law of Cosines is a fundamental trigonometric identity used to solve for unknown sides or angles in any triangle (whether right-angled or oblique). This law is beneficial when two sides and the included angle, or all three sides, are known.

Formula:

$$c^2 = a^2 + b^2 - 2ab \cdot \cos C$$

Where:

- $a, b,$ and c are the sides of the triangle,
- C is the angle between sides a and b.

Applications in Surveying:

- ### Determining Distances:

When working with oblique triangles, surveyors use the Law of Cosines to calculate unknown distances or angles, particularly when two sides and the included angle are known.

- **Triangulation**:

The Law of Cosines also calculates the distances between two survey points based on measured angles.

Distance, Distance Intersections

Distance intersections;

refer to methods used to determine the intersection point of two lines based on known distances.

Applications in Surveying:

- **Boundary Establishment**:

Surveyors can use distance and angle measurements to calculate the exact intersection of lines to mark boundaries.

- **Survey Control**:

Distance intersection calculations are vital in creating control points for larger surveying projects.

Interpolation

Interpolation estimates unknown values that fall between known values in a data set. In surveying, interpolation estimates distances, elevations, or other spatial data.

Key Concept:

- **Linear Interpolation**:

If you have two known points on a line, interpolation allows you to estimate a value at a point in between them.

Applications in Surveying:

- **Elevation Measurement**:

When surveyors have elevation data at two points, they use interpolation to estimate the elevation at a point in between.

The Compass Rule

The **Compass Rule** is a method used in land surveying to adjust measurements for minor errors in traverse data. It helps surveyors balance the closure error in a closed traverse by changing the survey angles.

Key Concept:

- **Adjustment of Angles**:

The Compass Rule applies a correction factor to the measured angles to ensure the calculated area or distance matches the actual boundary.

Applications in Surveying:

- **Traverse Adjustment**:

When measuring large parcels of land, the Compass Rule helps adjust minor errors in angle measurement to ensure the boundary is accurately surveyed.

Horizontal Curves

Horizontal curves refer to bends or changes in the direction of a road, railway, or boundary line in a horizontal plane. Surveyors use these curves to design roads, railways, and other infrastructure.

Key Concept:

- **Curve Radius**:

A curve's radius defines the bend's tightness. Surveyors use this to design roads or pathways safe for vehicles or people.

Applications in Surveying:

- **Road and Railway Design**:

Horizontal curves are essential in transportation design to ensure smooth turns for vehicles or trains.

Grades and Slopes

Grades and slopes describe the steepness or incline of a surface, particularly in roads, railways, and drainage systems.

Key Concept:

- **Grade**: The slope of a surface, usually expressed as a percentage (rise/run)

Formula:

$$\text{Grade} \, (\%) = \frac{\text{Vertical rise}}{\text{Horizontal run}} \times 100$$

Applications in Surveying:

- **Drainage Design**: Surveyors design roadways, railways, and drainage systems with specific grades to ensure proper drainage and smooth travel.

The Intersection of Two Grades

When two graded surfaces intersect, the surveyor must calculate the angle or alignment where they meet, ensuring proper alignment of roads, railways, or slopes.

Key Concept:

- **Intersection Angle**:

The angle at which two grades intersect is crucial for determining the design of the junction.

Applications in Surveying:

- **Road Design**:

Ensuring smooth transitions between different grades of roads or railways.

Vertical Curves

Vertical curves in a road or railway profile change the slope between two surface sections. These curves are critical for ensuring smooth transitions between different gradients.

Key Concept:

- **Curve Length**:

The length of the vertical curve affects the comfort and safety of transitions between different grades.

Applications in Surveying:

- **Road Design**:

Vertical curves are essential in highway and railway design to provide smooth elevation changes and improve driver safety.

A curve's radius defines the bend's tightness.

CHAPTER 7

COMMUNICATION AND REPORT WRITING SKILLS

Effective communication and report-writing skills are fundamental for quantity surveyors to accurately manage and convey technical, financial, and contractual information. These skills ensure smooth stakeholder interactions, proper documentation, and informed decision-making throughout project lifecycles.

Importance of Communication in Quantity Surveying

Communication is integral to quantity surveying. It facilitates coordination between clients, contractors, engineers, architects, suppliers, and regulatory authorities. Proper communication minimises misunderstandings, reduces disputes, and enhances project efficiency.

Key Components of Effective Communication

- **Clarity**

Information should be presented straightforwardly and free from ambiguity to ensure understanding by all stakeholders.

- **Consistency**

Regular updates are necessary to keep stakeholders informed about project progress and any deviations.

- **Accuracy**

The data, figures, and information shared must be precise to avoid costly errors.

- **Active Listening**

Engaging with stakeholders requires understanding their needs and responding effectively.

- **Adaptability**

The ability to adjust communication styles to suit the audience, whether technical experts or non-technical individuals.

- **Conflict Resolution**

Addressing disagreements or disputes diplomatically through negotiation and fact-based discussions.

Modes of Communication

1. **Verbal Communication**

Meetings, presentations, and on-site briefings.

2. **Written Communication**

Reports, emails, memos, and specifications.

3.   **Visual Communication**

Drawings, diagrams, schedules, and charts.

4.   **Digital Communication**

Using software and platforms for virtual meetings, collaborative tools, and document sharing.

Principles of Report Writing.

Reports are formal documents that present findings, evaluations, and recommendations in a structured format. Quantity surveyors use reports to inform stakeholders about costs, progress, risks, and performance metrics.

Essential Qualities of Effective Reports

- **Accuracy**.

Ensure data is factual, validated, and free from calculation errors.

- **Conciseness**.

Present the information briefly yet comprehensively, eliminating irrelevant details.

- **Logical Structure**

Organise content systematically for clarity and ease of navigation.

- **Objectivity**

Maintain neutrality, providing evidence-based findings and recommendations.

- **Professionalism**

Use formal language, avoid slang, and adhere to grammatical standards.

Types of Reports in Quantity Surveying

1. **Feasibility Reports**

Analyse project viability by assessing costs, timelines, and risks.

2. **Pre-Tender Estimates**

Provide cost estimates before tendering to guide budgeting.

3. **Tender Evaluation Reports**

Compare contractor bids based on compliance, costs, and capabilities.

4.   **Progress Reports**

Track ongoing work and assess whether milestones are being met.

5.   **Cost Reports**

Record financial performance during the project, including variations and contingencies.

6.   **Final Account Reports**

Summarise and reconcile all project expenditures with the original budget.

Report Writing Process

1. Planning

- Define the report's purpose, whether it is for cost analysis, tender evaluation, or progress tracking.

- Identify the target audience to determine the level of technical detail required.

- Gather supporting data, including drawings, cost estimates, and calculations.

2. Structuring the Report

1. **Title Page**

Includes the title, author, recipient, date, and purpose.

2. **Table of Contents**

Lists sections and page numbers for easy navigation.

3. **Executive Summary**

A brief overview of the findings and conclusions are provided.

4. **Introduction**

Explains the report's purpose, background, and scope.

5. **Methodology**

Describes data collection techniques, assumptions, and tools.

6. **Findings**

Presents analysed data supported by tables, graphs, or charts.

7. **Analysis and Discussion**

Interprets the findings, identifies trends, and highlights concerns.

8. **Conclusions and Recommendations**

Summarise the key outcomes and propose actions.

9. **Appendices**

Contains supplementary information, such as raw data or calculation sheets.

10. **References**

Cite any external sources used in the report preparation.

3. Reviewing the Report

- Verify numerical data and ensure calculations are accurate.

- Proofread for grammatical errors and structural inconsistencies.

- Confirm logical flow and alignment with the report's objectives.

- Obtain feedback from colleagues or supervisors before submission.

Tools and Techniques for Effective Report Writing

- **Word Processing Software**

Microsoft Word and Google Docs for drafting and formatting reports.

- **Spreadsheets**

Microsoft Excel for managing and analysing quantitative data.

- **Visualisation Tools**

Use charts, graphs, and tables to present data visually.

- **Project Management Software**

Platforms like Primavera or Microsoft Project to integrate cost and time schedules.

- **Templates**

Predefined templates standardise formatting and presentation.

- **Collaboration Tools**

Applications like Microsoft Teams and Google Workspace for shared editing and communication.

Common Communication Challenges in Quantity Surveying

Issues:

- **Technical Jargon**;

Misunderstandings may arise when non-technical stakeholders are unfamiliar with the terms used.

- **Language Barriers**;

Challenges in projects involving multinational teams.

- **Information Overload**

Excessive details can obscure critical points.

- **Misinterpretation**

Ambiguity in written or verbal instructions may lead to errors.

- **Timeliness**

Delayed reporting may affect project schedules and budgets.

Strategies for Overcoming Challenges:

- Simplify language to enhance understanding.

- Use visuals and diagrams to explain complex concepts.

- Encourage feedback and clarification to avoid assumptions.

- Set communication protocols for reporting and information flow.

- Provide training on cross-cultural communication for international projects.

Enhancing Communication and Report Writing Skills

- **Training and Workshops**;
Attend courses on professional writing, data presentation, and interpersonal communication.

- **Practice and Feedback**;

Develop sample reports and review them with peers or supervisors for improvements.

- **Reading Professional Reports**

Study examples from experienced surveyors to identify best practices.

- **Editing Tools**

Use grammar-checking software like Grammarly to refine language and style.

- **Continuous Learning**

Stay updated with industry trends and modern reporting techniques.

Practical Applications of Communication and Report Writing

- Preparing tender documents and cost estimates.

- Presenting findings to clients and management during progress meetings.

- Submitting payment applications and variation orders.

- Documenting disputes and preparing claims for arbitration or resolution.

- Recording daily logs and site observations during construction.

- Generating post-project reports to evaluate performance and outcomes.

MODULE TWO;

CHAPTER 8:

COST ESTIMATING AND ANALYSIS

1. Introduction to Cost Estimating

Cost estimating is forecasting the financial resources required for a construction project. It involves calculating direct and indirect costs associated with labour, materials, equipment, and other expenses. The goal is to predict project costs accurately to ensure feasibility and effective budgeting. Keywords: forecasting, financial resources, project feasibility, budgeting.

Cost Estimating and Analysis in Quantity Surveying

1. Introduction to Cost Estimating and Analysis

Cost estimating and analysis are critical components of the quantity surveying profession. They primarily focus on forecasting and evaluating the financial requirements of construction projects. Accurate cost estimating allows project stakeholders, including clients, contractors, and architects, to plan and budget effectively, reducing the likelihood of unforeseen costs and delays. On the other hand, cost analysis ensures that the project's financial management stays on track throughout its life cycle.

2. Cost Estimating Process

The cost estimating process involves several stages, from initial feasibility studies to the final cost projection. The key steps are:

- **Preliminary Estimate**:

This is the initial cost estimate made at the earliest stage of a project. It relies on limited information and focuses on providing a rough financial outline. It is used to determine the overall budget before detailed designs are made.

- **Detailed Estimate**:

This is prepared once more detailed information, such as architectural plans, technical specifications, and schedules, is available. The detailed estimate involves breaking the project into components and applying precise labour, materials, and overhead cost calculations.

- **Unit Rate Method**:

This method estimates costs based on established unit rates for materials, labour, and equipment. These rates are derived from historical data or industry standards. Each element of work is assigned a cost per unit, which is then multiplied by the quantities involved.

- **Bill of Quantities (BoQ)**:

A document that lists all materials, labour, and work items required for a construction project forms the basis for cost estimation. It provides detailed descriptions of the work and allows for calculating unit costs for each item.

3. Types of Cost Estimates

Different types of cost estimates are used depending on the stage of the project:

- **Conceptual Estimate**:

Often used in the early stages of a project when design details are minimal, this estimate provides an initial, broad overview of potential costs based on similar projects or historical data.

- **Order-of-Magnitude Estimate**:

Similar to conceptual estimates, order-of-magnitude estimates give a rough cost range. They are typically based on square footage or similar general measures.

- **Square Foot or Cubic Meter Estimate**: This method calculates costs based on the project size (e.g., price per square meter or cubic meter), making it easier to estimate costs when only general dimensions are known.

- **Detailed Quantity Take-Off (QTO) Estimate**:

This involves a more accurate calculation of material quantities, labour costs, and other similar costs based on the detailed drawings and specifications.

4. Cost Analysis

Cost analysis involves examining and interpreting cost data to determine a project's financial viability. The goal is to ensure the project remains within its budget and assess potential areas for cost reduction or optimisation.

- **Direct Costs vs. Indirect Costs**:

  o **Direct Costs**:

Costs directly attributed to the project's production, such as materials, labour, and equipment.

  o **Indirect Costs**:

Overhead costs that support the project but are not directly tied to any single component, including site management, administration, and utilities.

- **Cost Control and Monitoring**:

This process involves tracking the project's actual costs against the estimated costs to detect discrepancies. Key performance indicators (KPIs) such as cost variance (CV) and schedule variance (SV) are used to monitor financial performance.

- Life-cycle costing (LCC) is a method for evaluating the total cost of ownership of a project, including initial construction costs, operational costs, and maintenance expenses over the lifespan of the building or infrastructure. This helps in making informed decisions about the project's long-term economic impact.

5. Factors Affecting Cost Estimation

Several external and internal factors can influence the cost estimation process:

- **Market Conditions**:

Fluctuations in the prices of materials and labour can significantly affect estimates, particularly in volatile markets.

- **Design Changes**:

Any modifications made to the design after the estimate has been prepared can lead to cost overruns.

- **Location and Accessibility**:

The project's geographic location can influence material costs, labour availability, and transportation costs.

- **Project Scope**:

A clearly defined project scope helps provide a more accurate estimate. Scope creep—uncontrolled changes in the scope—can lead to cost increases.

- **Regulatory Requirements**:

Local building codes, environmental regulations, and other legal constraints can impact the cost of materials, labour, and time required to complete the project.

6. Methods for Cost Analysis

Cost analysis involves various techniques to refine estimates and assess overall project costs:

- **Cost-benefit analysis (CBA):**

This method evaluates a project's financial feasibility by comparing the anticipated costs against expected benefits. It helps decision-makers choose the most cost-effective project alternatives.

- **Break-Even Analysis:**

Considering all costs and revenues, this analysis helps determine when the project will start to generate profits. It is beneficial in projects with a revenue-generating component.

- **Earned Value Management (EVM)**

is a project management technique that integrates cost, schedule, and scope to assess project performance and progress. It compares the planned cost of work with the actual cost to date and the value of the work completed.

7. Challenges in Cost Estimating and Analysis

Despite being a core aspect of project management, cost estimating and analysis present various challenges:

- **Inaccurate Data**:

Without reliable, up-to-date data, estimates can be significantly off, leading to financial mismanagement.

- **Unforeseen Events**:

Delays caused by weather, material shortages, or unforeseen site conditions can impact the original cost estimates.

- **Complexity of Projects**: Larger, more complex projects may involve greater risk in estimating costs due to their multifaceted nature.

CHAPTER 9:

PROCUREMENT METHODS AND CONTRACT ADMINISTRATION

Procurement methods and contract administration are two fundamental areas in the construction industry that ensure a project is executed efficiently, within budget, and according to the specifications. Procurement involves acquiring the necessary services and materials for the project. At the same time, contract administration focuses on managing the construction contract during the project's execution to ensure compliance, monitor performance, and maintain accountability between stakeholders. Both areas are critical in preventing project delays, cost overruns, and disputes between parties.

Procurement Methods

Different procurement methods allow for varying degrees of control, risk, and cost distribution between the client and contractor. The chosen method should align with the project's scale, complexity, and specific requirements. The central procurement methods include:

a. Traditional Procurement (Design-Bid-Build)

- **Description**: In traditional procurement, the process is sequential, meaning the design is completed before construction begins. This method is typically preferred when the design is straightforward and there is a limited need for fast-track delivery. The client hires an architect or consultant to complete the designs and then invites contractors to tender the construction work. Once the contractor is selected, the project moves into the construction phase.

- **Advantages**:

  o The client has much control over the design phase and can make detailed design decisions before construction begins.

  o There is a clear distinction between the design and construction phases, simplifying roles and responsibilities.

- **Disadvantages**:

  o The time required for the sequential process can lead to delays in project completion.

○   If changes occur during construction, they can result in cost overruns and delays.

b. Design and Build (D&B)

- **Description**:

The design and build method streamlines the procurement process by combining the design and construction phases. In this method, a single contractor is responsible for designing and building the project. The client provides a general brief, and the contractor takes on both the design and construction responsibilities, often leading to quicker completion times.

- **Advantages**:

The method reduces project delivery times by allowing concurrent design and construction.

○   The client benefits from having a single point of contact for design and construction, reducing administrative overhead.

- **Disadvantages**:

o   The client has less influence over the design process once the contract is awarded.

o   Resolving issues with the design or construction quality can be more difficult due to the integrated nature of the contract.

c. Management Contracting

- **Description**:

Management contracting involves appointing a management contractor early in the design phase. The management contractor manages the trade contractors responsible for the construction work. The management contractor does not undertake the physical construction but ensures that all subcontractors are coordinated and the project progresses smoothly.

- **Advantages**:

The client can be involved early in the construction process, and the management contractor can ensure that work begins before all designs are finalised.

○   This approach can lead to faster project delivery as some work can commence while other parts are still being designed.

- **Disadvantages**:

Clients bear more financial risk, as they are often required to pay management contractor fees plus the costs of subcontractors.

○   Budgeting can be more challenging as costs are not fixed until later.

d. Construction Management.

- **Description**:

Construction management is similar to management contracting, but the construction manager acts as an agent for the client, coordinating the subcontractors directly. Unlike management contracting, the client has direct contracts with all the subcontractors, and the construction manager oversees their work.

- **Advantages**:

o   The client has more control over subcontractor contracts and can manage costs more effectively.

o   The project can be fast-tracked by allowing various construction parts to proceed as designs are finalised.

- **Disadvantages**:

The client assumes more responsibility for managing contracts, which can lead to increased administrative effort.

The complexity of handling multiple contracts can lead to coordination and project oversight challenges.

e. Turnkey Contracts

- **Description**:

In a turnkey contract, the contractor is responsible for all aspects of the project, from design to construction, and upon completion, delivers a fully functional building to the client. The client typically provides a broad set of requirements, and the contractor offers a finished, operational building without further input.

- **Advantages**:

The client only interacts with one entity, reducing management and coordination efforts.

The contractor ensures the project meets quality standards and is ready for use upon completion.

- **Disadvantages**:

The client has limited influence over the detailed design and construction process.

Changes or issues during the project can be harder to resolve because the contractor is solely responsible.

Contract Administration

Contract administration involves overseeing the execution of the construction contract to ensure compliance with all terms, including timelines, costs, and quality standards. Effective contract administration requires clear communication, monitoring progress, and addressing issues as they arise. It is crucial for preventing disputes and ensuring the project is completed as agreed.

a. Roles and Responsibilities in Contract Administration

- **Client**:

The client is responsible for providing funding, approvals, and support to ensure the project proceeds smoothly. They also review and approve design changes, variations, and other contract modifications.

- **Quantity Surveyor (QS):**

The QS plays a crucial role in the project's financial management. They prepare cost estimates, monitor the budget throughout construction, assess variations, and ensure the project remains within economic constraints. The QS also manages tendering processes and may help resolve disputes.

- **Contractor:**

The contractor is responsible for the actual construction work. They must comply with the contract's terms, including completing the job on time, within budget, and to the required quality standards.

- **Project Manager:**

The project manager oversees the project's day-to-day operations, ensuring that work progresses according to plan. They coordinate between the client, contractor, and other stakeholders to keep the project on track.

b. Contractual Documentation

- ### The Contract

the primary document establishes the terms of the agreement between the client and contractor. It includes key details such as the scope of work, delivery timelines, payment schedules, and penalties for non-compliance.

- ### Variation Orders:

If the scope of work changes, the client and contractor may issue a variation order, which formally changes the contract's original terms. Variations are common in construction projects and must be carefully managed to avoid disputes.

- ### Certificates of Payment:

The architect or QS issues these documents periodically to certify that the work completed to date aligns with the contract and warrants payment. They are vital in maintaining cash flow and ensuring timely payments to contractors.

c. Monitoring and Compliance.

- ### Progress Monitoring:

Regular inspections and reports ensure the project stays on schedule and within budget. By tracking work progress, costs, and schedules, stakeholders can identify issues early and take corrective actions when necessary.

- **Quality Control**:

Ensuring work meets the specified quality standards is a key responsibility during contract administration. It involves inspections, testing, and verification to ensure the end product meets design and safety standards.

- **Compliance with Regulations**:
Throughout the construction process, contractors must adhere to all relevant building codes, safety regulations, and environmental laws. Failure to comply can result in penalties or delays.

Common Contract Types

Construction projects use various types of contracts based on their specific requirements. Common contract types include:

- **Lump Sum Contract**:

The contractor agrees to complete the project for a fixed price. This type of contract is ideal when the scope of work is well-defined, minimising the risk of cost overruns.

- **Cost-Plus Contract**:

The contractor is reimbursed for actual costs incurred during construction, plus a fee for their profit. This type is proper when the project scope is uncertain but offers less cost certainty for the client.

- **Time and Materials Contract**:

The client agrees to pay for the contractor's time and materials used on the project. This contract is helpful for projects with undefined scopes but can lead to higher costs if not carefully managed.

CHAPTER 10

LEGAL ASPECTS OF SURVEYING AND CONSTRUCTION

Legal considerations in surveying and construction are critical to ensuring that projects are executed lawfully, ethically, and within the framework of applicable laws. Surveyors and construction professionals must adhere to various legal principles that govern property rights, contracts, safety regulations, environmental laws, and dispute resolution processes. Understanding these legal requirements is essential to mitigate risks, avoid legal disputes, and protect all parties involved in the construction process.

Key Legal Principles in Surveying and Construction.

a. Contract Law

Contract law forms the foundation of legal agreements in construction and surveying. Every construction project typically begins with a formal contract outlining the terms and conditions between the parties involved—clients, contractors, consultants, and suppliers.

146

Contracts are legally binding documents that specify each party's scope of work, timeline, payment schedules, and obligations. Any failure to meet these terms, whether delayed construction, substandard work, or non-payment, can lead to legal consequences. A well-structured contract helps prevent disputes and ensures all parties understand their roles and responsibilities.

b. Property Law

Property law governs the ownership and use of land, including the rights of individuals or organisations that own, lease, or hold interest in land for construction projects.

Understanding property law is crucial in construction, as the land where a project will take place must be legally available for development. Issues such as land title, zoning laws, and easements (legal rights to use a portion of land for a specific purpose) must be clearly defined to avoid future conflicts. In some cases, land disputes may arise, leading to delays or the cancellation of projects if property rights are not established.

c. Tort Law

Tort law addresses civil wrongdoings that harm individuals or property, and it is a significant aspect of construction law. Surveyors and contractors can be held liable for damage caused by negligence or failure to fulfil legal obligations.

Torts may include property damage, personal injury, or professional negligence in construction. For instance, if a surveyor makes an error in a land survey that leads to structural damage, they may be liable for negligence. Contractors must also avoid creating unsafe working conditions that could result in injuries or accidents, as they may be sued under tort law.

d. Construction Law

Construction law covers all legal issues related to the building process, from the initial planning stages to project completion. This area of law ensures that construction projects comply with safety standards, regulations, and contractual agreements.

Construction law addresses various issues, including permits, building codes, labour laws, safety standards, and dispute resolution. Adhering to construction law is vital for ensuring that the project is completed legally and that the construction process does not harm the environment or violate the rights of workers or the public.

Legal Documents in Surveying and Construction

a. Land Survey Reports

Survey reports are legal documents produced by surveyors that detail the boundaries, elevations, and characteristics of a piece of land or property. These reports serve as essential records in determining land ownership and development rights.

A land survey report is often used to settle property disputes, define boundaries, and support the issuance of title deeds. It is legally significant and is frequently referenced in legal proceedings related to property ownership or construction. Inaccurate or incomplete surveys can lead to legal challenges and may result in costly litigation.

b. Building Permits and Approvals

Obtaining the necessary permits and approvals is a legal requirement before construction begins. Local government authorities issue building permits and ensure the proposed construction project complies with zoning regulations, building codes, and other legal standards.

Construction projects cannot commence legally without the proper permits. This includes permits for excavation, foundation work, structural changes, and, in some cases, environmental assessments. Failure to obtain the required permits can result in fines, project delays, and even the need to demolish illegal constructions.

c. Construction Contracts

Construction contracts are written agreements between the client and contractor that outline the terms of the construction project, including project scope, cost, timeline, and dispute resolution procedures.

Construction contracts are legally binding, and both parties must adhere to the stipulated terms. Construction contracts often contain clauses related to warranties, guarantees, penalties for delay, payment schedules, and conditions for contract termination. A well-drafted contract can help avoid legal disputes by clarifying responsibilities and expectations.

Legal Responsibilities of Surveyors and Contractors

a. Duty of Care

Surveyors and contractors owe a duty of care to their clients, workers, and the public to ensure their actions or omissions do not cause harm. It includes performing work to a reasonable standard and taking necessary precautions.

Surveyors must accurately measure land boundaries and assess land suitability for development, while contractors are responsible for building projects according to specified plans and safety standards. If either party breaches this duty of care, they can be held liable for any resulting damages, including injury or property loss.

b. Compliance with Building Codes and Standards

Construction must comply with local and international building codes and standards, which ensure safety, environmental protection, and quality.

Building codes set out the minimum requirements for construction work, such as materials, structural integrity, and fire safety measures. Surveyors and contractors must ensure that all aspects of the construction adhere to these codes, and failure to comply can result in legal action or the project being halted.

c. Health and Safety Laws

Health and safety laws govern the working conditions of construction sites to prevent accidents, injuries, and fatalities. These laws apply to everyone involved in construction, from labourers to supervisors and visitors.

Contractors and surveyors are required to provide a safe working environment and comply with occupational health and safety regulations, including providing necessary personal protective equipment (PPE), ensuring proper site management, and addressing hazards. Legal actions may arise if workers are injured due to a failure to comply with these laws.

Dispute Resolution in Surveying and Construction

a. Arbitration

Arbitration is a standard method of resolving disputes in construction and surveying. It involves a neutral third party, an arbitrator, who listens to both sides of the conflict and makes a binding decision.

Arbitration is often preferred in construction contracts because it is quicker and more cost-effective than going to court. Construction contracts usually include arbitration clauses, which specify that disputes should be resolved through this method. While binding, arbitration can be more flexible than litigation, allowing for more specialised expertise.

b. Mediation

Mediation is a less formal dispute resolution process where an impartial mediator helps both parties reach a mutual agreement. Unlike arbitration, the mediator does not make binding decisions.

Mediation is often used in the early stages of a dispute to avoid lengthy and expensive legal proceedings. It allows parties to maintain control over the outcome and is especially useful when the relationship between the parties needs to be preserved.

c. Litigation

Litigation involves resolving disputes through the court system. It is the most formal and often the most expensive method of dispute resolution.

Construction disputes may end up in court when arbitration or mediation fails. Courts make binding decisions that must be followed, and litigation can be lengthy and costly. However, it may be necessary when the dispute is complex or other resolution methods have been exhausted.

Environmental and Regulatory Laws in Construction

a. Environmental Protection Laws

Environmental laws regulate the impact of construction projects on the environment, ensuring that developments minimise pollution, waste, and damage to ecosystems.

Construction projects must comply with regulations concerning waste disposal, water usage, and air quality. Surveyors and contractors must conduct environmental assessments and obtain necessary approvals before the project proceeds. Violations of environmental laws can lead to heavy fines, legal action, and reputational damage.

b. Zoning Laws and Land Use Regulations

Zoning laws dictate how the land can be used in specific areas, crucial for cities' and towns' legal and orderly development.

Before beginning construction, surveyors must ensure the land is zoned correctly for its intended use (e.g., residential, commercial, industrial). Construction projects that violate zoning laws can result in legal disputes, fines, or the cancellation of permits.

CHAPTER 11:

SUSTAINABLE BUILDING PRACTICES

Sustainable building practices, also known as green building, refer to the design, construction, and operation of buildings in a way that minimises their environmental impact. These practices focus on enhancing energy efficiency, reducing waste, and using environmentally friendly materials throughout a building's lifecycle. Quantity surveyors must be knowledgeable about sustainable building practices to assist clients in achieving ecologically responsible and cost-effective projects. The growing focus on sustainability in construction is driven by the need to reduce carbon footprints and contribute to long-term environmental conservation.

Principles of Sustainable Building Practices

a. Resource Efficiency

Resource efficiency is a key principle in sustainable building. It aims to minimise the use of natural resources throughout the building process. This includes efficiently using energy, water, and raw materials, which reduces waste and lowers operational costs.

By using resources efficiently, buildings can significantly reduce their environmental impact. For example, energy-efficient buildings use less energy for heating, cooling, and lighting, reducing utility costs and conserving energy resources. Water efficiency through rainwater harvesting or low-flow fixtures minimises the strain on local water supplies. Choosing materials with a lower environmental footprint, such as recycled or locally sourced materials, further enhances a building's sustainability.

b. Energy Efficiency

Energy efficiency involves designing buildings that use less energy while maintaining comfort and functionality. It can be achieved by selecting energy-efficient systems and technologies.

Buildings account for a significant portion of global energy consumption, making energy efficiency a crucial element of sustainable building practices. Strategies such as passive solar design, high-efficiency insulation, and energy-efficient HVAC systems and appliances can dramatically reduce a building's energy needs. Additionally, renewable energy sources like solar panels and wind turbines can be incorporated into the building's design to reduce dependency on non-renewable energy.

c. Minimizing Environmental Impact

Sustainable buildings are designed to minimise their negative impact on the environment. It includes reducing the carbon footprint, minimising pollution, and avoiding habitat destruction during construction.

Construction projects can significantly impact the environment through activities like site clearance, transportation of materials, and energy consumption. Sustainable building practices aim to mitigate these impacts by using environmentally friendly materials, recycling construction waste, and ensuring that buildings are energy-efficient. For example, selecting low-emission building materials helps reduce a structure's overall carbon footprint.

Key Elements of Sustainable Building Design

a. Building Materials

The choice of building materials plays a significant role in a construction project's sustainability. Sustainable materials are renewable, recyclable, locally sourced, and non-toxic.

Building materials can impact the environment long-term due to their production, transportation, and disposal. Using materials such as bamboo, recycled steel, or reclaimed wood helps reduce resource depletion. Low-emission materials like low-VOC (volatile organic compounds) paints and adhesives also improve indoor air quality. Furthermore, sourcing materials locally reduces transportation emissions and supports local economies.

b. Water Conservation

Water conservation is a critical aspect of sustainable building. It ensures that buildings use water resources efficiently while minimising waste.

Installing low-flow fixtures, efficient irrigation systems, and water recycling systems like greywater reuse can achieve water-efficient designs. In addition, rainwater harvesting systems can collect water from the roof for irrigation or non-potable purposes, reducing reliance on municipal water supplies. Incorporating these features conserves water and helps reduce utility costs over time.

c. Indoor Environmental Quality

Indoor environmental quality refers to the factors that affect building occupants' health, comfort, and well-being, such as air quality, lighting, and acoustics.

Sustainable buildings prioritise creating healthy indoor environments by using non-toxic materials, ensuring good ventilation, and maximising natural lighting. Proper indoor air quality can be achieved through air filtration systems and low-emission materials. Furthermore, acoustic design that reduces noise pollution is also part of creating a comfortable living or working environment.

Sustainable Building Certification and Standards

a. LEED (Leadership in Energy and Environmental Design)

LEED is a widely recognised green building certification program that provides a framework for identifying and implementing sustainable building design, construction, and operation practices.

The LEED rating system awards points based on various sustainability factors, such as energy efficiency, water usage, material selection, and indoor air quality. Projects that achieve specific points are awarded a LEED certification at one of several levels: Certified, Silver, Gold, or Platinum. LEED certification provides a clear benchmark for sustainability and helps attract environmentally conscious tenants and clients.

b. BREEAM (Building Research Establishment Environmental Assessment Method)

BREEAM

is another leading sustainability assessment method, primarily used in Europe, that evaluates the environmental performance of buildings based on energy, water, materials, and other factors.

Like LEED, BREEAM assesses a building's sustainability through various criteria, including energy usage, waste management, and the environmental impact of construction materials. BREEAM-certified buildings are recognised for their commitment to reducing environmental harm and improving sustainability throughout their lifecycle.

c. WELL Building Standard

The WELL Building Standard focuses on the health and well-being of building occupants, evaluating aspects like air, water, nourishment, light, fitness, and mental well-being.

While LEED and BREEAM focus on environmental sustainability, the WELL standard goes beyond that to consider how the built environment affects human health. Features such as access to natural light, proper ventilation, and spaces for physical activity are incorporated into WELL-certified buildings. This holistic approach improves occupant productivity, reduces absenteeism, and improves overall quality of life.

Advantages of Sustainable Building Practices

a. Cost Savings Over Time

While sustainable buildings may have higher upfront costs due to using energy-efficient systems and environmentally friendly materials, long-term operational savings can offset these initial expenses.

Energy-efficient buildings have lower utility bills, and water-saving measures can significantly reduce water consumption costs. Additionally, sustainable buildings are often more durable, reducing the need for costly repairs and replacements. Furthermore, buildings that meet high sustainability standards may qualify for tax incentives, grants, or reduced insurance premiums.

b. Improved Occupant Health and Comfort

Sustainable buildings prioritise the health and comfort of their occupants, which can lead to better employee productivity, higher tenant satisfaction, and fewer health-related absences.

Better air quality, access to natural light, and noise reduction contribute to a healthier and more comfortable living or working environment. It, in turn, can lead to increased productivity in office buildings or a better quality of life for residents in homes and apartments.

c. Environmental Benefits

The environmental benefits of sustainable building practices are numerous, including reduced carbon emissions, less pollution, and conservation of natural resources.

By utilising renewable energy sources, reducing waste, and using sustainable materials, green buildings contribute to the global effort to combat climate change. Reducing the environmental impact of construction helps preserve ecosystems, conserve biodiversity, and reduce the depletion of natural resources.

Challenges in Implementing Sustainable Building Practices

a. Initial Costs and Investment

The upfront costs associated with sustainable building practices, such as using eco-friendly materials or implementing energy-efficient technologies, can be higher than those associated with conventional construction methods.

While the long-term savings may outweigh these initial costs, the higher upfront investment can be a barrier for some developers and property owners. However, financial incentives like grants, tax rebates, and energy-saving programs can help offset these costs.

b. Lack of Awareness and Expertise

The construction industry, particularly in developing countries, may face challenges in implementing sustainable practices due to a lack of awareness or expertise in green building design.

Sustainable building practices require specialised knowledge and training from architects, engineers, contractors, and quantity surveyors. Implementing the best green building practices may not be easy without access to this expertise. Raising awareness and providing ongoing training is crucial for driving change in the industry.

CHAPTER 12:

BUILDING INFORMATION MODELLING (BIM)
APPLICATIONS

Building Information Modelling (BIM) is the
digital representation of a building's physical and
functional characteristics. BIM integrates multiple
aspects of a project into a single platform, which
allows for better visualisation, analysis, and
coordination throughout the building lifecycle. BIM
technology has revolutionised the construction
industry by enhancing collaboration, increasing
efficiency, and improving the overall quality of building
projects. Quantity surveyors use BIM to provide more
accurate cost estimations, ensure efficient use of
resources, and manage risks effectively.

Core Applications of BIM in Quantity Surveying

a.   Cost Estimation and Budgeting

- BIM allows quantity surveyors to
automatically generate detailed cost estimates from a
building's 3D model, ensuring that all aspects of the
construction are accounted for. Integrating cost data

with the design model helps provide more accurate and up-to-date cost predictions.

• Using BIM, quantity surveyors can analyse the cost of materials, labour, and time required for each project element. Automating cost estimation improves budget accuracy, minimises errors in manual calculations, and allows for better cost tracking throughout the project lifecycle.

b. Quantity Take-off and Measurement

• Traditional quantity takes off involves manual measurement and counting, which can be time-consuming and prone to errors. Conversely, BIM facilitates automated quantity take-off by extracting data directly from the 3D model.

• BIM enables quantity surveyors to obtain precise measurements of building components directly from the model, such as walls, windows, doors, and flooring. This reduces the need for rework and minimises discrepancies between the design and the final quantities, leading to more accurate and efficient procurement.

c. Clash Detection and Risk Management

- One of the key advantages of BIM is its ability to detect clashes and conflicts between different systems (e.g., structural, mechanical, and electrical systems) before construction begins. The model allows real-time collaboration so quantity surveyors and other professionals can identify and resolve issues early.

- By detecting potential issues such as overlapping pipes or structural interference, BIM minimises costly delays and rework during construction. This proactive risk management approach helps avoid budget overruns and schedule disruptions, ensuring smoother project execution.

Enhanced Project Coordination and Collaboration

a. Stakeholder Collaboration

- BIM enables seamless collaboration between all stakeholders involved in the construction process, including architects, engineers, contractors, and quantity surveyors. The centralised digital model allows all team members to work with the most up-to-date information in real-time.

- With BIM, changes made by one stakeholder are instantly reflected across the entire model, ensuring that everyone has access to the same data. This eliminates the need for multiple versions of plans and reduces the likelihood of miscommunication or mistakes.

b. Improved Communication

- The visual nature of BIM models enhances communication between all parties involved in a project. Complex designs can be easily visualised, making it easier for stakeholders to understand the project and its components.

- For quantity surveyors, BIM enhances their ability to explain cost implications, project timelines, and resource needs to clients, contractors, and other stakeholders. By presenting the project in 3D, they can illustrate how changes in design or materials impact the overall budget and schedule.

Project Scheduling and Time Management

a. 4D BIM for Scheduling

4D BIM integrates time as a fourth dimension into the model, linking construction activities to the building's 3D geometry. This allows quantity surveyors to generate precise construction schedules aligned with the building process.

- With 4D BIM, project teams can visualise the construction process in real time, ensuring that tasks are completed in the right sequence and within the designated timeframes. This helps manage delays, improve task sequencing, and optimise labour and resource allocation.

b. Timeline Simulation

- BIM software simulates the project timeline, allowing for better project planning and control. Quantity surveyors can analyse how changes in the project timeline affect costs and resource use.

- By simulating the construction process, BIM helps identify potential bottlenecks or inefficiencies in scheduling. This can lead to better resource allocation, improved site management, and faster project delivery.

Lifecycle Management and Facility Management

a. Building Lifecycle Information

- BIM provides a comprehensive view of a building's lifecycle, from design and construction to operation and demolition. This allows quantity surveyors to accurately predict maintenance and operational costs over the building's lifetime.

- By linking building components to their lifecycle data, such as expected lifespan and maintenance schedules, BIM helps plan future renovations or replacements. This ensures that the building remains cost-effective and operational throughout its lifecycle.

b. Asset Management

- Once construction is complete, BIM continues to be valuable for managing the building's assets. BIM assists in ongoing maintenance and repair activities by incorporating data on equipment, fixtures, and building systems.

- Facility managers can use BIM models to track the status and condition of building components, improving asset management and reducing downtime. For quantity surveyors, this information can be helpful in budgeting for future repairs and managing the building's ongoing costs.

-

Sustainability and Energy Efficiency

a. Energy Analysis and Simulation

- BIM models can be integrated with energy analysis software to simulate a building's energy performance. This allows for optimising the building's design to reduce energy consumption and improve efficiency.

- Using BIM for energy analysis enables quantity surveyors to evaluate different building systems, materials, and orientations to identify the most energy-efficient options. This lowers operational costs and contributes to environmental sustainability by reducing the building's carbon footprint.

b. Sustainable Material Selection

BIM aids in selecting sustainable materials by providing detailed information on material properties, environmental impacts, and lifecycle assessments. Quantity surveyors can use this data to choose materials that align with sustainability goals while staying within budget constraints.

- BIM helps analyse the environmental impact of construction materials and promotes the use of recyclable, low-impact materials. This encourages more sustainable building practices that contribute to overall environmental goals.

Benefits of BIM for Quantity Surveying

a. Increased Accuracy and Reduced Errors

- One of the most significant advantages of BIM for quantity surveyors is the increased accuracy of cost estimates and measurements. By automating the quantity take-off process, BIM reduces the chances of human error and ensures that data is consistent across all stakeholders.

- As a result, there is a reduction in the likelihood of discrepancies between the design and the actual quantities, leading to more accurate bids, cost estimates, and final project costs.

b. Cost Control and Budgeting

- BIM provides better control over project costs by enabling real-time tracking of costs, schedules, and resource allocation. Quantity surveyors can use BIM to adjust cost estimates as the project evolves, ensuring the budget remains on track.

- By utilising BIM's data-driven approach, project costs can be continuously monitored, which helps identify potential cost overruns early and allows for corrective actions to be taken promptly.

c. Improved Decision-Making

- BIM provides a wealth of data that allows quantity surveyors and other project stakeholders to make more informed decisions. Access to real-time information and simulations makes evaluating different design alternatives, material choices, and construction techniques easier.

• Visualizing the entire project lifecycle allows for better strategic planning, risk management, and decision-making, ultimately leading to more successful project outcomes.

Challenges and Limitations of BIM

a. High Initial Costs

• While BIM offers significant long-term benefits, the initial setup and implementation can be costly. This includes purchasing the necessary software, training staff, and converting traditional project processes into a BIM-enabled workflow.

• The high upfront costs may deter some smaller firms or project owners from adopting BIM, but over time, the cost savings, accuracy, and efficiency improvements typically outweigh the initial investment.

b. Learning Curve and Technical Expertise

• BIM requires a certain level of technical expertise, and the learning curve can be steep for individuals and organisations unfamiliar with the software. Training staff to effectively use BIM tools and integrate them into existing workflows is essential but time-consuming.

- Overcoming this barrier requires ongoing training and support, but the investment in building BIM expertise pays off in improved project delivery and management.

CHAPTER 13:

FINANCIAL MANAGEMENT IN CONSTRUCTION

Financial management in construction involves planning, monitoring, and controlling financial resources throughout a project's lifecycle. It is a crucial aspect of project management, ensuring that costs are controlled, budgets are adhered to, and economic risks are minimised. In construction, financial management includes handling costs, managing cash flow, securing funding, and maintaining profitability. Proper financial management helps ensure that the project stays within budget and is completed on time, essential for the construction project's success.

Cost Control and Budgeting

a. Cost Planning and Estimation

- The initial stage of financial management involves cost planning, which begins with cost estimation. This process includes estimating the direct and indirect costs associated with the project, such as labour, materials, overheads, and other expenses.

- Quantity surveyors play a critical role in cost estimation by analysing project designs and determining the required quantities of materials and labour. Accurate estimates are developed using historical data and industry standards to ensure the project budget is feasible and realistic.

b. Budgeting

- Once costs are estimated, a project budget is developed. The budget is a blueprint for the project and limits how much can be spent on various aspects of the construction.

- During construction, the budget is monitored and adjusted to ensure the project does not exceed financial constraints. Effective budgeting ensures sufficient funds for all aspects of the project, including unexpected costs, without causing financial strain.

Cash Flow Management

a. Cash Flow Forecasting

- Cash flow management ensures that the construction project has sufficient liquidity at each stage of its execution. Cash flow forecasting involves predicting the inflows and outflows of cash throughout the project lifecycle, considering payment schedules, material procurement, and labour costs.

- Cash flow forecasting is essential for preventing financial bottlenecks. By understanding when and where cash is required, financial managers can avoid delays caused by a lack of funds, ensuring the project progresses smoothly.

b. Managing Payment Schedules

- A key aspect of cash flow management in construction is managing payment schedules for contractors, subcontractors, and suppliers. Timely payments are crucial to maintaining good relationships with suppliers and workers and avoiding project delays.

- Payment schedules should be carefully structured to ensure funds are available when required. For example, payments are often linked to project milestones or deliverables. This helps keep cash flow stable and ensures contractors have the funds necessary to meet their obligations.

Financing the Construction Project

a. Securing Project Funding

- Construction projects often require significant funding, which can be sourced from various channels, such as bank loans, equity investments, or government grants. Financing is essential to cover initial expenses like purchasing materials, paying workers, and acquiring equipment.

- A comprehensive financial plan must demonstrate the project's viability to potential investors or lenders. The plan includes detailed cost estimates, projected revenues (in the case of commercial projects), and repayment timelines.

b. Risk Mitigation and Financial Contingency

- Securing financing also involves considering the financial risks associated with construction projects. Risks can include unforeseen site conditions, price inflation, or regulation changes. Financial managers must assess these risks and build contingencies into the project budget.

- A contingency fund is often set aside to cover unexpected costs. The project can continue smoothly despite unforeseen delays or increased costs by planning for potential financial setbacks.

Financial Reporting and Monitoring

a. Progress Reporting

- Financial monitoring involves regularly tracking the project's financial progress against the budget. This includes comparing actual costs with estimated costs and identifying any discrepancies. Regular reporting allows financial managers to adjust and keep the project on track.

- Progress reports typically include expenditures to date, forecasted costs, and reasons for any cost overruns. These reports help stakeholders make informed decisions and ensure the project stays within its financial targets.

b. Performance Metrics

- Performance metrics, such as earned value management (EVM), measure the project's financial health. EVM compares planned progress with actual progress in terms of time and cost. It helps financial managers assess whether the project is on track, ahead, or behind schedule and whether it is under or over budget.

- These performance metrics enable quantity surveyors and project managers to quickly identify potential financial issues and take corrective action before they escalate into significant problems.

Cost Control Techniques

a. Variance Analysis

Variance analysis involves comparing the project's actual and estimated costs to identify variances. These variances are categorised as favourable or unfavourable, depending on whether costs are under or over budget.

- By performing variance analysis, financial managers can identify areas where cost control is needed and take corrective action, such as revising project timelines or adjusting resource allocation to avoid further budget overruns.

b. Cost Benefit Analysis

• Cost-benefit analysis helps determine whether the anticipated benefits justify the financial investments in the project. For example, it is used to assess whether the cost of implementing a new technology or a change in design will result in enough savings or benefits to justify the expenditure.

• By evaluating the potential return on investment (ROI) for different project elements, financial managers can make decisions that contribute to the project's overall economic success.

Managing Subcontractor and Supplier Costs

a. Subcontractor Payments

• Subcontractors play a key role in construction projects, and their payments must be carefully managed to ensure that they are paid on time and that the project stays within budget. The contract should clearly define Payment terms with subcontractors, often linked to milestones or deliverables.

- Financial managers must ensure that subcontractor costs are accurately recorded and monitored. This helps prevent disputes over payments and controls the project's finances.

b. Supplier Management

- Construction projects rely heavily on suppliers for materials and equipment, and their costs must be carefully monitored. Supplier costs can fluctuate based on market conditions, such as material shortages or price increases.

- Financial managers must track material orders, delivery schedules, and suppliers' payments to ensure they remain within the project's budget. Maintaining good relationships with suppliers can also help negotiate better prices and favourable payment terms.

Financial Management Tools and Software

a. Construction Financial Management Software

• Modern financial management in construction is supported by specialised software designed to track costs, manage budgets, and generate reports. Software tools like Procore, Builder Trend, and Sage 300 Construction and Real Estate can integrate various aspects of project management and provide real-time financial data.

• These tools allow quantity surveyors and project managers to monitor the project's financial health, track payments, create cost reports, and perform variance analysis in one centralised system. They can also streamline administrative tasks, making financial management more efficient.

b. Spreadsheets and Financial Models

• For smaller projects or organisations, spreadsheets can help manage construction finances. Custom-built financial models in Excel can help track costs, manage cash flow, and perform fundamental economic analysis.

• Although spreadsheets are less robust than specialised software, they provide flexibility and are cost-effective for smaller firms or projects with less complex financial requirements.

Profitability and Financial Reporting for Stakeholders

a. Project Profitability Analysis

• Profitability analysis in construction helps determine whether the project is generating the anticipated financial returns. This includes calculating the gross profit margin, return on investment (ROI), and other profitability ratios.

• By regularly assessing the project's financial outcomes, financial managers can determine whether adjustments are needed to ensure that the project delivers a profit.

b. Reporting to Stakeholders

• Regular financial reporting to stakeholders, such as investors, contractors, and clients, ensures transparency and accountability. These reports typically include financial summaries, cost breakdowns, and performance indicators, enabling stakeholders to assess the project's economic health.

• Timely and accurate financial reports help build trust with stakeholders, ensuring their confidence in the project's financial management and progress.

CHAPTER 14:

INTERNSHIP AND INDUSTRIAL EXPOSURE

Internships and industrial exposure provide valuable real-world experience for students and professionals in Quantity Surveying. These opportunities allow individuals to apply the theoretical knowledge gained in academic settings to practical, on-the-ground projects. Industrial exposure helps students familiarise themselves with industry practices, enhance their skills, and prepare for future careers by working with experienced professionals.

The Role of Internships in Quantity Surveying

Quantity Surveying internships are structured opportunities for students or recent graduates to gain hands-on experience in the construction industry. Interns typically work under the supervision of senior quantity surveyors, assisting in various aspects of cost estimation, budgeting, and project management.

- **Skills Development**:

Internships help interns develop key skills such as cost analysis, preparing tender documents, managing project budgets, and understanding contract administration. These are essential skills for a successful career as a Quantity Surveyor.

- **Exposure to Industry Tools and Practices**:

Interns are exposed to industry-standard tools and practices, such as software for cost estimation and financial reporting. This exposure ensures they are well-versed in modern technologies and methodologies used in Quantity Surveying.

Types of Internship Programs

Internships can vary in terms of duration, structure, and responsibilities. Some common types include:

- Short-term Internships:

These typically last a few weeks or months and provide an introductory experience in Quantity Surveying. They focus on specific tasks such as cost estimation, taking off quantities, or assisting in preparing tenders.

- **Long-Term Internships**:

These programs last several months to a year and offer more in-depth exposure to multiple aspects of

Quantity Surveying. Interns may rotate through different departments or project stages, gaining a broader understanding of the industry.

- **Paid Internships**:

Some internships are paid, providing financial compensation to interns while offering them practical experience. These programs are typically more competitive and may provide additional learning opportunities.

Benefits of Internship and Industrial Exposure

Internships offer numerous benefits, both for the intern and for the industry.

- **Practical Application of Knowledge.**

: Interns can directly apply the concepts learned in the classroom to real-world projects, which helps bridge the gap between theory and practice. Working on live projects helps develop critical problem-solving skills.

- **Networking and Career Opportunities**:

Interns often build professional networks by interacting with industry professionals. These connections can lead to job offers, mentorship, or internship opportunities. Networking is essential in the construction industry, where personal relationships and referrals significantly influence career advancement.

- **Enhanced Employability**:

Internships enhance the employability of graduates by providing them with practical skills that employers highly value. Industry experience is often a critical factor when hiring new employees in the construction sector.

Key Areas of Exposure in Quantity Surveying

Internships in Quantity Surveying typically cover a wide range of activities that are essential for the role of a professional Quantity Surveyor. Some of the key areas of exposure include:

a. Cost Estimation and Budgeting

- Interns learn to prepare detailed cost estimates based on the project's scope, labour, materials, and other expenses. They may assist senior surveyors in calculating the quantities of materials and determining the cost per unit, helping to develop accurate budgets.

b. Tendering Process

- Interns are involved in preparing and submitting tender documents for construction projects. This includes reviewing tender specifications, analysing costs, and preparing project bids. Interns understand how to evaluate bids and select contractors based on cost-effectiveness.

c. Project Management

- During their internship, students may assist in managing various aspects of construction projects, such as overseeing the financials, ensuring project timelines are met, and ensuring that costs remain within the allocated budget. They learn to handle day-to-day operations and administrative tasks related to project management.

d. Contract Administration

- Interns learn to manage and administer construction contracts, including understanding terms and conditions, payment schedules, and legal obligations. This exposure is crucial for understanding construction projects' commercial and legal aspects.

e. Quantity Take-off and Site Measurements

- Interns often perform quantity take-offs, calculating the materials and labour required for a project. Site measurements and ensuring accuracy are critical skills that interns develop through their industrial exposure.

Challenges and Learning Opportunities

Internships provide various learning opportunities, but they also come with challenges.

- **Adaptability**:

Interns may face challenges adjusting to the pace and complexity of real-world projects. However, these challenges offer learning opportunities to develop adaptability and problem-solving skills, which are essential for success in the construction industry.

- **Time Management**:

Balancing multiple tasks in a project requires effective time management. Interns learn to prioritise tasks, meet deadlines, and work efficiently under pressure, which are vital skills for a Quantity Surveyor.

- **Industry-Specific Knowledge**:

Interns are exposed to industry-specific challenges such as fluctuating material prices, regulatory compliance, and the environmental impact of construction projects. These real-world issues provide valuable insights that can be difficult to grasp in an academic setting fully.

Importance of Mentorship in Internship Programs

Mentorship is a critical aspect of successful internships in Quantity Surveying. Interns are typically paired with experienced professionals who provide guidance, share industry knowledge and offer feedback on performance. A strong mentor-mentee relationship can:

- **Enhance Skill Development**:

Mentors can provide valuable insight into specific technical aspects of Quantity Surveying and advise on best practices.

- **Foster Professional Growth**:

Mentorship helps interns develop a deeper understanding of the industry and offers career advice. This is especially important for those looking to specialise in areas such as cost management, project management, or contract administration.

Evaluating the Success of Internship Programs

To ensure the success of an internship program, both the intern and the hosting organisation must assess the outcome. Evaluations typically focus on the following:

- **Skills Gained**:

Interns and their mentors should assess the skills acquired during the internship, such as cost estimation, budgeting, or contract management.

- **Career Progression**:

Internships are an excellent way to evaluate the intern's potential. Employers may offer full-time positions to interns who have demonstrated strong performance, work ethic, and relevant skills.

- **Feedback from Interns**:

Interns need to provide feedback on their experience, including the challenges they faced, the learning opportunities offered, and the program's overall effectiveness. This feedback helps organisations improve their internship programs.

MODULE THREE

CHAPTER 15: ADVANCED MEASUREMENT AND ESTIMATION.

Introduction

Estimating determines the quantities and the expected costs for a specific task or project. If the available funds are insufficient for the estimated cost, the work may be completed partially, scaled down, or modified in terms of specifications. To prepare an estimate, the following requirements are essential:

1.   Drawings, including plans, elevations, and sections of key areas.

2.   Detailed specifications regarding the artistry quality and materials' properties.

3.   The standard schedule of rates for the current year.

Units of Measurement

Units of measurement are categorised based on their nature, shape, size, and their role in determining payments to the contractor. The basic principles for measurement units typically include:

a) Individual items such as doors, windows, and trusses are counted in numbers.

b) Works involving linear measurements, such as cornices, fences, handrails, and bands of a specific width, are measured in running meters (RM).

c) Works that involve surface areas, like plastering, whitewashing, and partitions of specified thickness, are measured in square meters (m²).

d) Works involving volumes, such as earthworks, cement concrete, and masonry, are measured in cubic meters (m³).

Measurement Rules

The guidelines for measuring each item are typically outlined in IS-1200. However, some general rules are listed below:

1.   Measurements should be taken for the completed work, and the description of each item should encompass materials, transportation, labour, tools, equipment, and all overheads necessary to complete the job to the required specifications, size, and shape.

2.   the order should follow the length, width, and height or thickness sequence in measurement.

3.   All work should be measured according to the following tolerances:

i) Linear measurements should be rounded to the nearest 0.01m.

ii) Areas should be rounded to the nearest 0.01 sq.m.

iii) Volume measurements should be rounded to the nearest 0.01 cubic meter.

4.   Similar types of work under different conditions or natures should be measured separately under distinct items.

5.   The bill of quantities should thoroughly outline the materials, proportions, and artistry and accurately represent the work to be done.

6.   For masonry (stone or brick) or structural concrete, categories should be measured separately, with heights described as follows:

a) From foundation to plinth level

b) From the plinth level to the first-floor level

c) From the first floor to the second-floor level, and so on.

Requirements of Estimation and Costing

1.   An estimate provides an idea of the project's cost and helps assess its feasibility,

determining whether it can be executed within the available funds.

2.   The estimate helps determine the time required to complete the work.

3.   An estimate is essential for inviting tenders, quotations, and arranging contracts.

4.   The estimate is used to control costs during the execution phase.

5.   An estimate ensures that the proposed plan aligns with the available funding.

Procedure of Estimating

Estimating involves the following steps:

1.   Preparing a detailed estimate.

2.   Calculating the rate for each unit of work.

3.   Preparing an abstract of the estimate.

Data Required to Prepare an Estimate

To prepare an estimate, the following data is needed:

1.   Drawings (plans, elevations, sections, etc.)

2.   Specifications.

3.   Rates.

Drawings

Preparing an estimate becomes problematic if the drawings are unclear or lack complete dimensions. It is crucial to ensure that drawings are complete before estimating.

Specifications

- **General Specifications**: These outline the nature, quality, class, and materials used in various parts of the work, offering a broad idea of the building's construction.

- **Detailed Specifications**: These provide in-depth details about each work item, specifying the quantities and quality of materials, proportions, preparation methods, artistry, and execution procedures.

Rates

The unit rates for each work item are required to prepare an estimate. This includes:

1.   The unit rates for each item of work.

2.   The rates for various materials used in construction.

3.   The cost of transporting materials.

4.   The labour wages for skilled and unskilled workers, including masons, carpenters, and other tradespeople.

Complete Estimate

Many people assume that the estimate for a structure only covers the cost of land, materials, and labour. However, numerous other direct and indirect costs are involved, which should be considered and included in the final estimate.

Lumpsum

When preparing an estimate, there are cases where it is not feasible to work out every detail, especially for certain types of work. These are lump sum (L.S.) items, which refer to tasks where the cost cannot be precisely measured or broken down into individual components. These items are grouped and given a single estimated cost.

Lumpsum items in an estimate may include:

1.   Water supply and sanitation systems.

2.   Electrical installations include meters, motors, and other electrical components.

3.   Architectural features and decorative elements.

4.   Contingencies and unforeseen items.

Generally, a percentage of the total estimated cost is allocated for these lump sum items. Even if sub-estimates are prepared or the work is completed, the actual price should not exceed the lumpsum amount set aside in the central estimate.

Work Charged Establishment

Many temporary workers, including skilled supervisors, assistants, and watchmen, are employed during construction. Their salaries are drawn from the lump sum designated to establish the work charged. This refers to the establishment costs directly tied to the job. Typically, 1.5% to 2% of the estimated cost is set aside for the work-charged establishment.

Methods of Taking Out Quantities

Quantities such as earthwork, foundation concrete, brickwork in the plinth, and superstructure can be calculated using various methods, including:
a) Longwall - short wall method
b) Centre line method
c) Partly use the centre line and short wall method.

Long Wall-Short Wall Method

In this method, the wall running along the room's length is called the "long wall," while the perpendicular wall is the "short wall." To determine the length of these walls, start by calculating the centre line lengths of each wall. The length of the long wall is calculated by adding half of the breadth at each end to the centre line length. For the short wall, the length is determined by subtracting half of the breadth from the centre line length at each end. The quantities are then derived by multiplying these lengths by the breadth and depth.

Centre Line Method

This method is effective for walls of similar cross-sections. It involves multiplying the total centre line length by the breadth and depth of the respective items to get the total quantity. In cases where cross walls or partitions join the main wall, half the breadth for each junction reduces the centre line length. Careful consideration is given to these junctions to ensure accurate measurement. Estimates prepared using the centre-line method are often quick and precise.

Partly Centre Line and Partly Cross Wall Method

This method is used when external walls are one thickness and internal walls differ. In such cases, the centre line method is applied to external walls, while the long wall-short wall method is used for internal walls. This approach is especially suited for buildings with varying wall thicknesses and foundation levels. Engineering departments widely practice it due to its flexibility.

Detailed Estimate

A detailed estimate involves calculating the quantities for various work items and determining each cost. This process is carried out in two stages:

1.   **Details of Measurements and Calculation of Quantities.**

The work is divided into individual items such as earthwork, concreting, brickwork, and plastering. Measurements are taken from the drawings and recorded in the appropriate columns of a predefined format.

Types of Estimates

Abstract of Estimate Form

A detailed estimate should include:

1. Report

2. Specifications

3. Drawings (plans, elevations, sections)

4. Design charts and calculations

5. Standard schedule of rates

Factors to Consider When Preparing a Detailed Estimate

1. **Quantity and Transportation of Materials**:
For large projects, material requirements are higher. Bulk materials are often purchased and transported at lower rates.

2. **Location of Site**:
The site should be chosen to minimise the risk of

damage during transit, loading, unloading, and storing materials.

3.   **Local Labour Charges**:

The skills, suitability, and wages of local labour are considered when preparing the detailed estimate.

Data

The process of calculating each item's cost or rate per unit is known as Data. To prepare this, rates for materials and labour are derived from current standard schedules, while the quantities of materials and labour required for one unit of work are obtained from a Standard Data Book.

Fixing the Rate Per Unit of an Item

The rate per unit of an item includes the following:

1.   **Material Quantity and Cost**:

Material requirements are based on the Standard Data Book (S.D.B.), including costs like first, freight, insurance, and transportation charges.

2.   **Labour Cost**:

The labour cost is determined by multiplying the

number of labourers required for each unit of work by their daily wage.

3.   **Cost of Equipment (T&P)**:
Special equipment, tools, and plants may be needed for specific tasks. In these cases, 1 to 2% of the estimated cost is provided for equipment.

4.   **Overhead Charges**:
Overhead costs such as office rent, equipment depreciation, staff salaries, postage, and lighting are around 4% of the estimated price.

Methods for Preparing Approximate Estimates

Preliminary or approximate estimates are required to study various aspects of a project and obtain administrative approval. They help determine whether the net income from a commercial project justifies the investment. The approximate estimate is based on practical knowledge and the costs of similar projects. The forecast includes a report explaining the project's necessity and utility and a site or layout plan. A contingency allowance of 5 to 10% is included. Methods used to prepare approximate estimates include:

a) Plinth Area Method

b) Cubical Contents Method

c) Unit Base Method.

Plinth Area Method

The construction cost is calculated by multiplying the plinth area by the plinth area rate. The area is determined by multiplying the building's length and breadth (outer dimensions). Factors such as material quality, labour type, foundation type, building height, roof type, woodwork, fixtures, and the number of stories determine the plinth area rate. According to IS 3861-1966, the following areas are included in the plinth area calculation:

1.   Area of walls at floor level

2.   Internal shafts for sanitary installations (up to 2.0 m²), lifts, air-conditioning ducts, etc.

3.   Area of bars at terrace level (a covered space on a terrace roof used for shelter during the rainy season)

4.   Non-cantilever type porches

Areas not included in the plinth area calculation:

1.   Area of lofts

2.   Unenclosed balconies

3.   Architectural features like bands, cornices, etc.

4.   Domes, towers above terrace level

5.   Box louvres and vertical sun breakers.

Cubical Contents Method

This method is generally used for multi-story buildings and is more accurate than the plinth area and unit base methods. The cost of the structure is estimated by multiplying the total cubical contents (volume) of the building by the local cubic rate. The volume is calculated by multiplying the building's length, breadth, and height, measuring from wall to wall (excluding the plinth offset). Costs for string courses, cornices, corbelling, etc., are ignored.

Revision Task

1.   What is the difference between preliminary, detailed, supplementary, and revised estimates?

2.   What do you understand by the following terms?

a. Overhead cost

b. Analysis of rates

c. Contingencies and supervision charges

d. Standard measurements book

, e. Prime cost

f. Provision of tools and plants and work-charged establishments in an estimate

g. Lump-sum items

3.   Distinguish clearly between the following:

a. Revised estimate vs. supplementary estimate

b. Administrative approval vs. technical sanction

c. Plinth area estimate vs cube rate estimated. Contingencies vs supervision charges

, e. Preliminary estimate vs. detailed estimate

4.   Explain the following terms:

a. Schedule of rates

b. Cube rate estimate

c. Preliminary estimate

d. Provisional items

, e. Carpet area

f. Revised estimate

g. Contingencies

h. Book value

i. Prime cost

j. Floor area

5.   Discuss the advantages and disadvantages of the cubic meter method of approximate estimation.

6.   Explain the following terms:

a. Record drawings

b. Standard measurement book

7.   What conditions must be fulfilled before starting work?

8.   What do you understand by detailed and general specifications?

9.   When and where are the following estimates used?

a. Annual repair estimate

b. Revised estimate

c. Supplementary estimate

10.   What is meant by a preliminary estimate?

11.   What documents should accompany a preliminary estimate?

12.   What are the different types of estimates?

13.   Which method provides the exact cost, and why?

14.   What are the different sections of a report for a building estimate?

15.   Explain the term 'provisional items.'

16.   Given the following data, prepare a preliminary estimate of a four-story office building with a total carpet area of 2000 sq. M to obtain the government's administrative approval. Corridors, verandas, lavatories, staircases, etc., may take up 40% of the built-up area.

- The plinth area rate is Rs. 1325 per sq.m.

- Extra for special architectural treatment: 0.5% of building cost.

- The extra cost due to the deeper foundation at the site is 1% of the building cost.

- Extra for water supply and sanitary installation: 8% of building cost.

- Extra for internal electrical installation: 12.5% of the building cost.

- Extra for other services: 5% of building cost.

- Contingencies: 2.5%

- Supervision charges: 10%

Answer:

To prepare a preliminary estimate for the four-story office building, we will follow these steps:

Step 1: Calculation of Built-Up Area

The total carpet area of the building is given as 2000 sq. However, 40% of the built-up area will be occupied by corridors, verandas, lavatories, staircases, etc. Therefore, we first need to determine the total built-up area:

$$\text{Built-Up Area} = \frac{\text{Carpet Area}}{\left(1 - \text{Percentage taken by non-usable area}\right)}$$

$$\text{Built-Up Area} = \frac{2000}{\left(1 - 0.40\right)} = \frac{2000}{0.60} = 3333.33 \, \text{sq.m.}$$

Step 2: Calculation of Basic Building Cost.

The plinth area rate is Rs. 1325 per sq.m. To calculate the essential building cost, we multiply the built-up area by the plinth area rate:

$$\text{Basic Building Cost} = \text{Built-Up Area} \times \text{Plinth Area Rate}$$

$$\text{Basic Building Cost} = 3333.33 \times 1325 = 4,420,833.25 \, \text{Rs.}$$

Step 3: Calculation of Additional Costs

1. **Special Architectural Treatment:** 0.5% of the building cost

   $$\text{Extra for Special Architectural Treatment} = 0.005 \times 4,420,833.25 = 22,104.17 \, \text{Rs.}$$

2. **Deeper Foundation at Site:** 1% of the building cost

   $$\text{Extra for Deeper Foundation} = 0.01 \times 4,420,833.25 = 44,208.33 \, \text{Rs.}$$

3. **Water Supply and Sanitary Installation:** 8% of the building cost

   $$\text{Extra for Water Supply and Sanitary Installation} = 0.08 \times 4,420,833.25 = 353,666.66 \, \text{Rs.}$$

4. **Internal Electrical Installation:** 12.5% of the building cost

   $$\text{Extra for Internal Electrical Installation} = 0.125 \times 4,420,833.25 = 552,604.16 \, \text{Rs.}$$

5. **Other Services:** 5% of the building cost

Step 4: Calculation of Contingencies and Supervision Charges

1. Contingencies: 2.5% of the building cost

$$\text{Contingencies} = 0.025 \times 4,420,833.25 = 110,520.83\,\text{Rs.}$$

2. Supervision Charges: 10% of the building cost

$$\text{Supervision Charges} = 0.10 \times 4,420,833.25 = 442,083.33\,\text{Rs.}$$

Step 5: Total Estimate= summation of

- Basic Building Cost
- Extra for Special Architectural Treatment
- Extra for Deeper Foundation
- Extra for Water Supply and Sanitary Installation
- Extra for Internal Electrical Installation
- Extra for Other Services
- Contingencies
- Supervision Charges

Total Estimate=

4,420,833.25+22,104.17+44,208.33+353,666.66+552,604.16+221,041.66+110,520.83+442,083.33=

5,167,062.66Rs

CHAPTER 16:

DISPUTE RESOLUTION AND ARBITRATION

Disputes are common in quantity surveying and construction projects due to various factors such as contractual disagreements, delays, cost overruns, or differing interpretations of the project scope. Effective dispute resolution is crucial to maintaining project timelines and budgets and ensuring all parties are satisfied with the outcome. Arbitration is one of the primary methods used to resolve disputes, especially when litigation is not an option or preferred option.

Types of Disputes in Quantity Surveying

1.   **Contractual Disputes**

o   These arise when there is a disagreement over the terms and conditions of the contract, including payment terms, deliverables, scope of work, and timelines. Such disputes can lead to delays and cost implications if not resolved promptly.

o   The role of a quantity surveyor here is vital in clarifying terms related to pricing, project timelines, and deliverables, thereby preventing potential disputes through careful contract management.

2.   **Cost Disputes**

Cost disputes often occur when there is a conflict over the final cost of a project. Disagreements may stem from variations, material costs, or labour charges that differ from initial estimates.

Accurate cost estimation and regular monitoring of project expenses are essential to prevent such disputes, as quantity surveyors play a key role in tracking costs and variations.

3.   **Quality and Standards Disputes**

Disagreements can also arise over the materials' quality, the artistry, or whether the finished work meets the agreed standards.

    o   Regular inspections, quality control measures, and adherence to agreed specifications can help avoid these disputes, with quantity surveyors ensuring compliance with construction standards.

4.   **Delays and Time-related Disputes**

construction delays are a common source of conflict, particularly when they lead to penalties or extended project timelines. These delays may arise from unforeseen circumstances, lack of resources, or poor project management.

Understanding time-related clauses and efficient project scheduling can mitigate the risk of delays-related disputes.

Dispute Resolution Methods;

1.   Negotiation

o   Negotiation is the first step in resolving any dispute and involves direct communication between the parties involved to reach a mutual agreement. This approach is informal and relies on the goodwill of the parties to resolve issues.

o   The quantity surveyor may act as an intermediary in negotiations, providing objective assessments of costs, project scope, and timelines to help both parties reach an amicable resolution.

2.   Mediation

o   Mediation is a process where an impartial third party helps facilitate discussions between disputing parties to assist in finding a resolution. Unlike arbitration, mediators do not make decisions but instead work to guide the parties toward a mutually agreeable solution.

- ○ Mediation can be a cost-effective and quicker alternative to litigation, with the mediator often being an expert in the field, such as a senior quantity surveyor or construction expert.

3. Adjudication

- ○ Adjudication is often used in construction disputes and is a quicker, more informal process than arbitration or litigation. An adjudicator is appointed to make a temporary, binding decision on the dispute, with the possibility of appeal through arbitration or litigation if the decision is unsatisfactory.

- ○ This method is beneficial in ensuring that the project continues without prolonged delays, as the adjudicator's decision is usually issued within a set time frame, often 28 days.

4. Arbitration

- ○ Arbitration is a formal dispute resolution process where an impartial third party, known as the arbitrator, makes a binding decision on the dispute after hearing both sides. It is often preferred over litigation because it is less time-consuming and more private.

○　　　Arbitration can resolve disputes regarding costs, delays, or other contractual issues. The arbitrator's decision is final, and while there may be limited grounds for appeal, arbitration offers a relatively efficient way to resolve complex disputes.

Arbitration Process

1.　**Arbitration Agreement**

○　　　The arbitration process begins when both parties agree to resolve the dispute through arbitration. This agreement is often stipulated in the contract, with a clause outlining the steps for arbitration in a dispute.

○　　　The agreement may include selecting an arbitrator, the location of the arbitration, and the procedural rules to be followed during the arbitration process.

2.　**Appointment of Arbitrator**

○　　　The parties involved usually agree on an arbitrator, or the court may appoint one if they cannot agree. The arbitrator should be impartial and have expertise in the field relevant to the dispute, such as construction or quantity surveying.

o   The arbitrator needs to maintain neutrality to ensure fairness in the process and uphold the integrity of the dispute resolution.

3.   **Hearing and Presentation of Evidence**

o   During the arbitration hearing, both parties present their case by submitting relevant evidence, including documents, witness testimony, and expert opinions. The arbitrator reviews all information and listens to arguments before making a decision.

o   Unlike in court cases, the arbitration process is generally less formal, allowing for a more streamlined presentation of facts and evidence.

4.   **Award and Enforcement**

o   After considering all the evidence and arguments, the arbitrator issues an award, a binding decision on the dispute. This award may include directives for payment, changes to work, or other remedies.

o   The court can enforce the award if one party refuses to comply with the decision. However, because the arbitration process is binding, most parties comply with the award to avoid additional legal proceedings.

Advantages of Arbitration

1.   Faster Resolution

o   Arbitration is generally quicker than litigation, with many disputes being resolved in months rather than years. The process is designed to be more efficient, reducing the burden of prolonged disputes on the project.

2.   Cost-Effectiveness

o   Arbitration can be less expensive than traditional court procedures due to its more streamlined process. Legal fees, court costs, and the time spent on the case are typically lower in arbitration.

3.   Confidentiality

o   Unlike public court cases, arbitration is confidential, ensuring that sensitive information and business dealings remain private. This is particularly beneficial for companies concerned with protecting their reputation.

4.   Expert Arbitrators

o   Arbitrators are often experts in a specific field related to the dispute, such as construction or quantity surveying. Their specialised knowledge

223

ensures that the decision is based on a thorough understanding of the industry.

CHAPTER 17:

ADVANCED SOFTWARE FOR QUANTITY SURVEYING (AUTOCAD, REVIT)

In the modern construction industry, quantity surveyors rely on advanced software tools to streamline processes, enhance accuracy, and improve efficiency. AutoCAD and Revit are two of the most widely used software programs for quantity surveying, each offering distinct features that cater to different aspects of construction planning, design, and management. These tools facilitate tasks ranging from detailed design and modelling to accurate quantity take-offs and cost estimation.

AutoCAD in Quantity Surveying

AutoCAD is one of the most widely used design and drafting software programs in construction. It is commonly used by quantity surveyors, architects, and engineers to produce precise drawings and designs for construction projects.

1.   **Drafting and Design**

AutoCAD allows users to create detailed 2D and 3D designs that serve as the foundation for project development. These drawings include plans, elevations, and sections, essential for accurate measurements and quantity take-offs.

o   Quantity surveyors use AutoCAD drawings to extract quantities directly from the drawings, making it easier to generate accurate material lists and cost estimates.

2.   Quantity Take-offs

With the help of AutoCAD's specialised tools, quantity surveyors can extract quantities from the digital drawings using the "Data Extraction" function. This process automates the take-off of quantities like areas, lengths, and volumes from 2D or 3D models.

Quantity accuracy is greatly improved, reducing the chances of human error in manual calculations. This automation saves time and enhances the accuracy of cost estimations.

3.   Integration with Other Software

o   AutoCAD is often integrated with software like Excel and specialised quantity surveying tools. This integration allows for seamless data transfer between a project's design and estimating stages, enhancing overall project management.

o   For instance, AutoCAD drawings can be linked with spreadsheet software to produce cost estimates based on accurate quantities derived from the drawings.

4.   **Customization and Flexibility**

○   AutoCAD is highly customisable, allowing quantity surveyors to tailor the software to their needs. Users can create templates, standardised drawing sets, and custom tool palettes that simplify repetitive tasks.

○   Customization helps improve workflow efficiency by allowing surveyors to focus on project-specific needs rather than redoing tasks each time.

Revit in Quantity Surveying

Revit is a building information modelling (BIM) software developed by Autodesk. It offers more advanced features than AutoCAD, particularly for quantity surveying tasks. Revit is designed to handle all aspects of the construction process, from initial design to detailed construction and maintenance.

1.   **Building Information Modelling (BIM)**

o   Revit uses BIM to create detailed 3D models that include not only the physical characteristics of the building but also its functional and operational properties. These models enable quantity surveyors to perform more detailed analyses of the project's design and construction.

o   By using BIM, quantity surveyors can perform real-time updates to quantities as the design changes, ensuring that cost estimates and schedules remain accurate throughout the project lifecycle.

2.   **Quantity Take-offs and Material Schedules**

o   One of Revit's most significant advantages is its ability to automatically generate accurate quantity take-offs and material schedules directly from the 3D model. The software can calculate areas, volumes, lengths, and quantities of building materials like concrete, steel, and finishes.

o   These take-offs can be instantly updated whenever changes are made to the design, reducing the likelihood of errors and discrepancies in the final cost estimation.

3.   **Integration with Estimating and Costing Tools**

o   Revit integrates seamlessly with estimating software like CostX, Bluebeam, or Sage Estimating. This integration helps bridge the gap between design, quantity take-offs, and cost estimating, allowing for more accurate and efficient financial planning.

o   With these integrations, quantity surveyors can link Revit models directly to cost databases, ensuring that the materials listed in the schedule are accounted for with current market prices and labour rates.

4.   **Collaboration and Coordination**

o   Revit's cloud-based capabilities enhance collaboration between quantity surveyors, architects, engineers, and contractors. Since all parties work with the same model, changes are visible to all stakeholders in real time, preventing miscommunication and reducing the chance of costly errors.

o   The collaborative nature of Revit makes it especially useful for large-scale projects where multiple disciplines are involved, and close coordination is necessary.

5.   **Visualization and Conflict Detection**

o   Revit's advanced visualisation tools allow quantity surveyors to view 3D models of the project, helping them understand complex structures and detect any design conflicts or errors early on. This is crucial for ensuring the construction process runs smoothly and within budget.

o   Revit can highlight potential issues, such as clashes between different building systems (e.g., electrical, plumbing, HVAC) and resolve them before construction begins, helping avoid costly redesigns or rework during construction.

Key Differences Between AutoCAD and Revit in Quantity Surveying

While AutoCAD and Revit are essential tools in quantity surveying, they differ in their approach and functionality.

1.   **2D vs. 3D Modelling**

o   AutoCAD primarily focuses on 2D design and drafting, though it has limited 3D capabilities. This makes it suitable for projects that require detailed drawings but don't need complex modelling.

o   Revit, on the other hand, is fully BIM-enabled. It specialises in 3D modelling and is particularly useful for large, intricate projects requiring detailed analysis and real-time updates.

2.   **Accuracy and Efficiency**

o   AutoCAD requires manual input and calculations for quantity take-offs, making it more prone to human error. While the software is versatile, it does not offer the same level of automation as Revit.

o   Revit's automatic updates and quantity extraction from the 3D model significantly improve accuracy and efficiency. Any changes made to the model are instantly reflected in the quantity schedules, reducing the need for rework.

3.   **Collaboration**

o   AutoCAD is less collaborative compared to Revit. While AutoCAD files can be shared between team members, coordinating changes is manual and may lead to discrepancies.

o   Revit enhances collaboration with its cloud-based features and real-time updates and ensures all project stakeholders use the latest design information.

Benefits of Using Advanced Software in Quantity Surveying

1.   **Improved Accuracy**

o   Both AutoCAD and Revit help improve the accuracy of quantity take-offs and cost estimates by reducing human error and automating many manual tasks.

o   Automated quantity extraction ensures that the correct measurements are taken directly from the drawings or models, reducing discrepancies between what was estimated and what is required.

2.   **Time Efficiency**

o   Using AutoCAD and Revit significantly reduces the time spent on manual calculations and drawing revisions. This allows quantity surveyors to focus on higher-value tasks like cost optimisation and risk management.

○　　　Revit's real-time updates make it easier to track changes in the design and adjust the estimates, accordingly, saving time and effort.

3.　　**Better Collaboration**

○　　　The cloud-based features and the integration capabilities of both AutoCAD and Revit with other software tools foster better collaboration among team members. This ensures that all disciplines are aligned and reduces the likelihood of errors due to miscommunication.

○　　　Enhanced communication also leads to smoother project execution and the ability to address problems as they arise, preventing costly delays and disputes.

CHAPTER 18:

FINAL RESEARCH PROJECT

A final research project in quantity surveying is a comprehensive academic endeavour where students or professionals undertake in-depth research on a specific topic related to the field. This project helps to demonstrate a thorough understanding of the concepts, methodologies, and practices in quantity surveying while contributing new knowledge to the profession. The research process involves defining the research problem, collecting data, analysing it, and presenting findings that can influence the industry.

1. Identifying the Research Topic

The first step in any research project is selecting a relevant and feasible topic. In quantity surveying, the topic should address current challenges, new methodologies, or emerging trends in the field. Topics can range from cost estimation techniques to construction sustainability or technology's impact on quantity surveying practices.

- **Relevance**:

The chosen topic should contribute to solving real-world problems within the quantity surveying profession, such as improving cost estimation accuracy, reducing project delays, or enhancing project management efficiency.

- **Feasibility**:

 Feasibility considerations include available resources, time constraints, and access to data. A clear, focused research question is essential to guide the research process effectively.

2. Research Objectives and Hypotheses

Once a topic is selected, the next step is to define the research objectives and hypotheses. Research objectives outline the study's goals, while the hypothesis presents a testable proposition that the researcher aims to prove or disprove.

- **Research Objectives**:

These are the specific aims or goals of the study. For instance, a research objective might be to assess the effectiveness of a new cost estimation software in reducing project costs.

- **Hypothesis**:

A hypothesis is a statement the research will attempt to validate. For example, "The implementation of BIM technology reduces the time required for quantity take-offs in construction projects."

3. Literature Review

The literature review is essential to any research project. It provides a comprehensive overview of existing knowledge on the research topic. It involves reviewing academic papers, industry reports, books, and other credible sources to understand previous research findings and identify gaps in knowledge.

- **Purpose**:

The review helps contextualise the research within the broader academic and professional landscape. It also allows the researcher to identify methodologies used in similar studies and assess their effectiveness.

- **Methodology**:

A thorough review of existing literature involves analysing trends, controversies, and the evolution of thought in quantity surveying. It positions the new research within the ongoing conversation in the field.

4. Research Methodology

The research methodology outlines the techniques and approaches for gathering and analysing data. In quantity surveying, qualitative and quantitative methods can be used depending on the nature of the research question.

- **Qualitative Methods**:

 These may involve case studies, interviews, and focus groups to understand subjective factors, such as perceptions of cost estimation accuracy or project management strategies.

- **Quantitative Methods**:

These methods focus on numerical data and often involve surveys, experiments, or statistical analysis to evaluate patterns and trends. For example, regression analysis can be used to study the correlation between labour costs and project delays.

- **Sampling**:

Depending on the research design, researchers may use random, stratified, or purposive sampling to select participants or data points for the study. Ensuring a representative sample is key for the validity of the findings.

5. Data Collection

Data collection is gathering information that will be analysed in the research project. The method of collection depends on the research design and the available resources.

- **Primary Data**:

Primary data is collected directly from sources through surveys, interviews, observations, or experiments. This data is typically more reliable and tailored to the specific research objectives.

- **Secondary Data**:

Secondary data comes from existing sources like academic journals, industry reports, historical records, or government publications. It helps to support primary data or provides background information for the study.

6. Data Analysis

Data analysis involves examining and interpreting the collected data to uncover patterns, relationships, or trends that address the research objectives or hypothesis. This stage is critical for transforming raw data into meaningful conclusions.

- **Qualitative Analysis**:

In qualitative research, analysis involves coding the data into themes or categories and interpreting response patterns. For example, in interviews with quantity surveyors, researchers may analyse the data to identify common challenges in estimating project costs.

- **Quantitative Analysis**: Quantitative data is typically analysed using statistical tools such as SPSS, Excel, or specialised software for regression analysis, correlation studies, or hypothesis testing. This allows the researcher to draw conclusions based on statistical significance.

-

7. Findings and Discussion

In this section, the researcher presents the results of the data analysis and discusses the implications of the findings about the research objectives and hypotheses.

- **Results**:

The findings should be presented clearly and logically, using tables, charts, and graphs to illustrate key points. For example, if the research is about the impact of BIM on cost savings, the findings might

include comparing costs before and after BIM
implementation.

- **Discussion**:

This section interprets the results, exploring how they
align with or contradict previous research. It provides
insights into the significance of the findings and
suggests how they could be applied in quantity
surveying.

8. Conclusion and Recommendations

The conclusion summarises the main findings
of the research and offers recommendations based on
the study. It also reflects on the limitations of the
research and suggests areas for future study.

- **Conclusion**:

This section encapsulates the key insights gained
from the research, answering the research question or
validating the hypothesis. It provides a succinct
overview of the study's contributions to the field.

- **Recommendations**:

Based on the findings, the researcher may suggest
practical solutions or strategies for quantity surveyors,
such as adopting certain technologies, improving cost

estimation practices, or refining project management methods.

9. Project Presentation and Defence

Once the research is completed, the findings are often presented in a formal report or thesis. In many cases, the researcher must defend their work through a presentation, often in front of an academic panel or industry experts.

- **Presentation**:

The researcher presents the key findings, methodologies, and conclusions, answering questions from the panel and addressing any challenges or limitations in the research.

- **Defence**:

The defence involves justifying the research design, methodology, and findings. It is an opportunity to showcase the depth of knowledge and expertise developed during the research process.

10. References and Citations

Credible sources must support a final research project, and all references should be appropriately cited to avoid plagiarism. Researchers should use a consistent citation style, such as APA, MLA, or Chicago, to credit the original authors of the works referenced throughout the research.

- **Credibility**:

Proper referencing strengthens the validity of the research and ensures its academic integrity. The reference list should include all sources consulted during the research process, including books, articles, and reports.

11. Final Report Writing

The final report is a comprehensive document that presents the research in a structured and coherent manner. It includes an introduction, literature review, research methodology, data analysis, findings, conclusion, and references.

- **Structure**:

The report should follow a clear, logical structure, with each section clearly labelled and organised. The use of headings and subheadings helps readers navigate the report quickly.

- **Clarity**:

The writing should be clear and concise, avoiding unnecessary jargon while conveying complex ideas effectively. Visual aids like graphs, tables, and charts are often included to enhance understanding.

Glossary

Quantity Surveying

commonly used terms in Quantity Surveying will help you navigate the complex construction world. Construction projects are complex enough without wondering what a particular word or term means.

Adjudication:

Adjudication is a dispute resolution process for settling disputes between parties in a construction contract. It involves an adjudicator, an independent third party with expertise in the construction industry. Adjudication provides a quick and cost-effective means of resolving disputes, allowing construction projects to move forward without lengthy delays.

AFP

Application for Payment refers to a request for interim or complete payment by a contractor or subcontractor in the construction industry.

Accrual

Value of work completed but not yet charged

Benchmarking:

Comparing a project's costs and performance to industry standards or similar projects to identify areas for improvement. By benchmarking, construction professionals can gain insights into best practices and identify potential inefficiency areas or cost-saving opportunities. This information can be used to make informed decisions and optimise project performance.

Bill of Quantities (BoQ):

A detailed document that lists and describes all the materials, labour, and other costs required for a construction project forms the basis for tendering and cost control. The BoQ provides a comprehensive breakdown of the project's components, enabling accurate cost estimation and comparing contractor bids during the tendering process.

BOQ Coding:

A systematic method of labelling items in the Bill of Quantities to enhance organisation and identification. By assigning specific codes to each item in the BoQ, quantity surveyors and other construction professionals can quickly locate and refer to particular items. This improves efficiency and reduces the chances of errors or confusion during the construction process.

BOQ Rate:

The cost associated with a unit of measurement in the Bill of Quantities is often expressed as a rate per unit (e.g., price per square meter). BOQ rates are crucial for accurately estimating project costs. They reflect the pricing of labour, materials, and other resources required for each unit of measurement specified in the BoQ.

Bonds

"Protects the client against the non-performance of a contractor."

BoQ

Bill of Quantities;

A document used in tendering in the construction industry in which materials, parts, and labour are itemised

BCIS

Building Cost Information Service of the Royal Institution of Chartered Surveyors (RICS)

described by RICS as 'the leading provider of cost and price information to the construction industry and anyone else who needs comprehensive, accurate and independent data.

Capital Expenditure (CapEx):

Funds are allocated for long-term asset investments, such as buildings or equipment. Capital expenditure plays a vital role in the construction industry, as it allows for acquiring and developing assets that generate revenue or provide essential infrastructure for various projects.

Capital Works:

Capital works projects involve constructing, improving, or extending physical assets and infrastructure. They encompass many construction activities, such as building new structures, renovating existing ones, or developing infrastructure like roads, bridges, and utilities. These projects contribute to the overall development and enhancement of communities and economies.

Cash Allowances:

A budgeted amount for specific items or tasks that may not be fully defined in the BoQ. Cash allowances provide flexibility in accounting for costs that may vary or are subject to uncertainty. They allow for adjustments during the construction process to cover unforeseen expenses or changes in design or specifications.

Cash Flow Forecast:

A projection of the expected inflows and outflows of funds throughout a construction project. A cash flow forecast helps construction professionals understand the timing of cash inflows (e.g., payments from clients) and outflows (e.g., payments to suppliers and contractors). It assists in managing the project's financial resources effectively and ensures sufficient funds are available to support ongoing construction activities.

Cash Flow Management;

is the process of monitoring and optimising the flow of funds throughout a construction project to meet financial obligations. Effective cash flow management is vital for successfully executing a construction project. It involves tracking incoming and outgoing payments, managing invoicing and collections, and making strategic decisions to ensure a steady and balanced cash flow.

Change Order:

Change orders are written documents that formalise project scope, schedule, or cost changes. They are a standard part of the construction process, allowing for modifications or adjustments to the original plans as the project progresses. They provide clarity and transparency regarding any changes, ensuring all parties know and agree to the revised terms.

Construction Phase Plan:

A safety and logistics plan outlining the key elements of a construction project, including site layout, access, and health and safety measures. The construction phase plan is essential for managing a project's health and safety. It helps identify potential hazards, establish safety protocols, and ensure compliance with relevant regulations and standards.

Contingency Fund:

An allocated budget for unforeseen or unexpected expenses may arise during construction. Contingency funds are a buffer to cover unexpected costs not accounted for in the initial project budget. By setting aside a contingency fund, construction professionals are prepared to address unforeseen challenges without compromising the project's progress or quality.

Cost Control

involves managing and monitoring project costs to ensure they align with the budget and making necessary adjustments. It also involves ongoing tracking of expenses, comparing them with the budgeted costs, and implementing measures to address any deviations or potential cost overruns. Effective cost control is essential for project profitability and completion within the allocated budget.

Cost Documentation

is the systematic recording and organisation of project cost information for reference and analysis. It includes records of expenses, invoices, receipts, and other financial data related to the construction project. Proper documentation allows for accurate cost estimation, monitoring, and analysis, helping construction professionals make informed decisions and improve cost management practices.

Cost Estimation

is the process of calculating the anticipated cost of a construction project, often using historical data and cost analysis. It is vital for determining the viability and feasibility of a construction project. It involves assessing the required resources, considering market conditions, factoring in labour and material costs, and accounting for potential risks and contingencies.

Cost Forecasting:

Cost forecasting is predicting future project costs based on current and historical data. It helps construction professionals anticipate and plan for a project's financial implications. It assists in efficient resource allocation, financial planning, and risk management, allowing for timely and informed decision-making.

Cost Index:

A numerical value adjusts historical cost data for inflation or market fluctuations. Cost indices provide a benchmark for measuring changes in cost levels over time. By applying appropriate cost indices, construction professionals can accurately compare historical and current costs and make meaningful cost projections.

Cost Overrun:

Cost overruns are when the actual project costs exceed the budgeted or estimated costs. They can occur due to unforeseen changes, scope creep, material price fluctuations, or delays. Managing cost overruns is crucial for maintaining project profitability and ensuring financial viability.

Cost Plan:

A comprehensive budgeting tool used to allocate costs to different stages of a construction project, the cost plan breaks down the overall project budget into specific categories and items. It is a reference document for tracking and controlling project costs, ensuring that resources are allocated efficiently and following project requirements.

Cost Variance:

The numerical difference between the budgeted or estimated cost and the actual cost of a project. Cost variance indicates how project costs deviate from the initial budget or estimate. Positive cost variance suggests cost savings, while negative cost variance indicates cost overruns.

Cost-to-Complete (CTC):

An estimate of the remaining costs required to complete a construction project. The cost-to-complete is calculated based on the work completed and the anticipated expenses of remaining activities and resources. It provides valuable insights into the project's financial status and helps manage resources and cash flow.

Continuous Professional Development

"Defined as a commitment for people who are chartered, or going through the process of becoming chartered, to update their skills and knowledge to remain professionally competent continually. At Metroun, we have an educational platform called Metroun Learning, which offers over 30 hours of formal CPD.

Cut and Fill

"A process in earthwork where the amount of material removed (cut) roughly equals the amount added (fill)."

Contingency Allowance

"A budgetary provision made to cover unforeseen costs in a construction project."

Change Order

"a document used to alter the original agreement on a construction project."

CVR

Cost Value Reconciliation – Gives you an ongoing account of a contract's profitability by measuring cost against value at different points in a contract's lifecycle, right through to completion.

Credit note

Reduces the amount owed on the invoice document

Cost Code – used to separate costs into specific categories

Daywork

"Work that cannot be priced in the usual way and is paid for daily, based on time and materials used."

Defects Liability Period

"A specific timeframe after the completion of construction during which the contractor is obliged to rectify any defects that arise"

Depreciation;

is the decrease in the value of assets over time, often considered in life cycle costing. Depreciation considers the wear and tear, obsolescence, and ageing of assets used in construction projects. By factoring in depreciation, construction professionals can determine the expected lifespan of assets and allocate costs accordingly.

Dilapidation Survey:

An assessment of the condition of neighbouring buildings and infrastructure before construction to document their state and protect against potential damage claims. Dilapidation surveys involve a detailed inspection and documentation of the existing structures and infrastructure adjacent to a construction site. This helps establish a baseline condition and can be used as evidence to support claims of alleged damage during construction.

Detailed Design;

the phase where the design is refined and plans, specifications and estimates are created

NEC, JCT & FIDIC

These are different types of construction contract providers. See our playlists for more details.

ECI – Early Contractor Involvement

act of involving the principal contractor during the early stages of design for professional input

Earned Value Analysis (EVA):

Earned value analysis assesses a project's performance by comparing the budgeted work cost to the actual price. It provides insights into a construction project's efficiency and progress. Comparing the value of completed work against the associated expenses helps identify potential deviations from the initial plan and allows for proactive management and corrective actions.

Earned Value Management (EVM):

Earned value management is a project management technique that integrates cost, schedule, and performance data to assess project progress and predict outcomes. It enables construction professionals to monitor and control project performance by measuring the value of completed work against the planned budget and schedule. Earned value management offers a comprehensive view of project health and facilitates data-driven decision-making.

Estimating Software:

Computer programs designed to assist quantity surveyors in accurately estimating project costs and generating BoQs. Estimating software eliminates manual calculations and provides efficient tools for cost estimation, quantity take-off, and generating detailed BoQs. These software solutions are invaluable for enhancing accuracy, consistency, and productivity in the estimation process.

Earned Value Analysis

"A project management technique objectively measuring project performance and progress, considering scope, time and cost."

Extension of Time (EOT)

"A mechanism by which a contractor requests a longer period than contractually agreed to complete construction".

Feasibility Study:

An initial analysis to assess whether a construction project is economically viable and worth pursuing. Feasibility studies evaluate various aspects, such as financial, technical, legal, and environmental, to determine the project's feasibility. These studies help stakeholders make informed decisions and mitigate risks before committing substantial resources to a construction project.

Final Account:

The final account is a financial statement that records all costs incurred and payments made throughout a project. It is used to close out the project financially. The final account provides a comprehensive overview of all financial transactions related to the construction project. It includes a summary of costs, payments, and any adjustments or variations made during the project's lifecycle.

Float

The total extra time beyond what the contractor needs, which is typically included at the end of its construction programme

Forecast

Anticipated future spending on a project based on evidence and/or assumptions

Front-end loading (FEL)

is an early project phase that involves comprehensive planning, feasibility studies, and initial cost estimation. It aims to establish a solid foundation for a construction project by conducting thorough research, defining project objectives, and assessing potential risks and constraints. This phase sets the stage for effective project management, cost control, and successful project delivery.

GMP

Guaranteed Maximum Price

"A contract in which the contractor is reimbursed for actual costs plus fee up to a ceiling price, which is the maximum amount the client will pay."

Inflation Rate:

Inflation is the annual percentage increase in the general price level, which affects project costs. Inflation rates impact the purchasing power of money over time and can significantly influence project budgets and cost projections. Construction professionals consider inflation rates when estimating project costs to adjust for the potential price increase during the project's duration.

Interim Payment:

Partial payments are made to contractors at various project stages, usually based on completed and certified work. Interim payments provide financial support to contractors and suppliers throughout the construction process. They help maintain cash flow and incentivise timely completion of project milestones while ensuring that the contractor's financial needs are met.

Life Cycle Costing:

Life cycle costing is an analysis that considers the total cost of a project over its entire lifespan, including construction, maintenance, and operational costs. It considers the initial investment and the costs associated with ongoing maintenance, repairs, and eventual demolition or disposal. This approach enables stakeholders to make informed decisions by evaluating the long-term financial implications of the project.

Liquidated Damages:

A contractor must pay the client a predetermined amount of compensation for delays or non-performance as specified in the contract. Liquidated damages protect the client financially if the contractor breaches a contract. They provide an agreed-upon measure of compensation for potential losses incurred due to project delays or substandard performance.

LAD

Liquidated Ascertained Damages

"An agreed rate of damages paid by the contractor to the employer for a particular breach of contract"

Lump Sum Contract

"A contract in which the contractor agrees to complete the work for a fixed price."

Liability;

In its broadest sense, 'liability' refers to a responsibility placed on someone or something that places them at a disadvantage. For example, responsibility for the design or responsibility to pay a company money

Material Requisition:

A document used to request the purchase of specific materials for the construction project. Material requisitions provide detailed information about the required materials, including quantities, specifications, and delivery instructions. This document helps streamline the procurement process, ensuring the necessary materials are sourced and delivered promptly.

Notification

Official form of communication used between parties to notify the other of items such as payment, queries, warnings, change, defects, extension of time & final accounts.

Operating Expenditure (OpEx):

Funds are allocated for day-to-day operational expenses, including maintenance and utilities. Operating expenditures cover the ongoing costs associated with the maintenance and operation of buildings, equipment, and infrastructure. Proper budgeting and control of OpEx are essential for efficient construction project management and the sustainability of the built environment.

Overhead and Profit

"Costs associated with running a business typically applied as a fixed percentage of the actual cost or price of works."

Performance Bond:

A contractor provides a financial guarantee to ensure the completion of a project according to the contract terms. Performance bonds protect the client's interests by ensuring the contractor will fulfil their contractual obligations. Should the contractor fail to deliver as agreed, the performance bond can cover the costs of completing the project or remedying any deficiencies.

Procurement Strategy:

A procurement strategy outlines how materials and services will be sourced and acquired for a construction project, including traditional procurement, design-and-build, or construction management methods. The strategy considers project requirements, budgetary constraints, and resource availability. It helps construction professionals make informed decisions regarding a project's most suitable procurement approach.

Progress Payment:

Payments are made to the contractor at specified intervals as work progresses, based on the project's completion of milestones or stages. Progress payments provide contractors with regular cash flow while allowing clients to verify the satisfactory completion of specific project milestones. These payments help ensure a fair and balanced financial arrangement that aligns with the project's progress.

Project Closeout:

The final phase of a construction project includes inspections, documentation, and the project handover to the client. Project closeout involves completing any outstanding work, conducting final inspections, verifying compliance with specifications, and obtaining necessary approvals or certifications. It also includes preparing project documentation and handing the client relevant information, warranties, and keys.

Prime Cost Sum

"An allowance in the contract for the supply of work or materials to be provided by a nominated subcontractor or supplier."

Provisional Sum

"Funds set aside for specific elements of works which are not yet defined enough to produce an accurate price."

Quantity Breakdown

"A detailed analysis of quantities for each element of a project."

Quantity Surveyor (QS):

A professional who manages and controls costs within the construction and building industry. They are responsible for estimating, procurement, and cost management of construction projects. Quantity surveyors ensure that projects are delivered within budgetary constraints, managing costs effectively and providing accurate financial forecasts.

Quantity Surveying Software:

Software tools aid in managing project costs, tendering, and financial analysis. Quantity surveying software enables professionals to streamline and automate tasks such as quantity take-off, cost estimation, and generating detailed reports. These software solutions help improve accuracy, efficiency, and collaboration in quantity surveying.

Quantity Surveyor's Daywork Rates:

Rates are used to calculate the cost of additional work not covered in the original contract, often based on the quantity surveyor's day work rates. Daywork rates provide a standardised method for pricing additional works or variations that may arise during construction. These rates ensure transparency and fairness in the cost assessment of extra work.

Quantity Surveyor's Report:

A detailed document summarising cost analysis, project progress, and recommendations for clients and stakeholders. The quantity surveyor's report provides a comprehensive overview of the project's financial aspects, including cost breakdowns, payment summaries, and progress evaluations. This report helps stakeholders assess project performance, make informed decisions, and ensure effective cost management.

Quantity Surveyor's Standard Method of Measurement (SMM7):

A set of guidelines and rules for quantifying building works, materials, and labour. The quantity surveyor's SMM7 provides a standardised framework for accurately measuring and quantifying the components of construction projects. This method ensures consistency and enables meaningful comparisons across different projects, facilitating accurate cost estimation and control.

Quantity Take off:

Quantity take-off is quantifying and itemising the materials and labour required for a construction project based on the BoQ. It involves identifying, measuring, and calculating quantities for each item specified in the BoQ. This process provides the foundation for accurate cost estimation, procurement, and resource planning.

Retention

A percentage (usually up to 5% of the contract sum) of each payment made under a construction contract is withheld to ensure that the work is completed to the required standard.

RICS, CICES & CIOB –

These are different types of charter institutions

Rate Analysis:

Rate analysis is the process of analysing unit rates in the Bill of Quantities to determine the cost of specific items or activities. It involves reviewing and comparing rates provided in the BoQ to assess their reasonableness and accuracy. This analysis helps validate the cost estimates and contributes to effective cost control and financial management throughout the construction project.

Retention:

The client retains a portion of the contract until the contractor completes their obligations and rectifies any defects. Retention ensures the contractor fulfils their contractual obligations and completes the project to the required standards. It provides financial security for the client and can be released to the contractor upon satisfactory completion of the project.

Retention Bond:

A financial guarantee that can replace the retention amount to ensure a contractor's performance. A retention bond provides an alternative to retaining a portion of the contract sum. It is a financial instrument the contractor provides to the client, guaranteeing their obligations and assuring them that any defects or incomplete work will be rectified within a specified period.

Risk Management:

Identifying, assessing, and mitigating potential risks and uncertainties affecting the project's cost and schedule. Risk management involves proactive measures to anticipate and minimise the impact of risks on the construction project. By identifying and addressing potential risks early on, construction professionals can reduce the likelihood of cost overruns, delays, and other adverse consequences.

Risk Register:

A risk register is a document that lists and assesses potential project risks and proposed mitigation strategies. It is a centralised reference for all identified risks, their likelihood, potential impacts, and the actions planned to mitigate or manage them. Monitoring and addressing potential risks throughout the project lifecycle helps construction professionals stay proactive and prepared.

Risk Register

"A vital tool in project management, the Risk Register is a document that lists potential risks, their impact, and mitigation strategies."

Re-measurement

"The process of measuring the actual work done on-site may differ from initial estimates."

Scope Creep:

Scope creep is the uncontrolled expansion of a project's scope, leading to increased costs and extended timelines. It occurs when the project's requirements gradually increase beyond the initially agreed-upon scope. Effective scope management is essential for controlling costs and ensuring the project remains focused and aligned with the client's objectives.

Sunk Cost:

Sunk costs are expenditures that have already been incurred and cannot be recovered. They are often considered in project cost analysis. Sunk costs are expenses spent on a project that cannot be recovered, regardless of the project's outcome. When analysing the project's financial viability, it is crucial to focus on future costs and benefits rather than sunk costs.

Subcontractor:

A company or individual hired by the main contractor to perform specific tasks or supply materials for a construction project. Subcontractors play a significant role in the construction industry, contributing their expertise and specialised services to complete various aspects of a project. They often work under the direction of the main contractor, providing specific skills or resources required for specific project phases.

Substantial Completion:

Substantial completion is the stage of a construction project where the work is almost finished, allowing for occupancy or beneficial use. It signifies that the project is functional and can be occupied or utilised for its intended purpose, although some minor works may still need to be completed. This milestone is a significant achievement and may trigger contractual obligations and payment milestones.

Snagging

"Snagging involves identifying and listing defects or unfinished work in a construction project for rectification."

Subcontractor

"A person or company hired by the main contractor to perform a specific task as part of the overall project."

Scope/ Works Information;

a general description of the work expected under a particular contract.

Specification

A well-structured, detailed description of the quality, standards, artistry, materials, and completion of work to be done which evolves across a project

Tender

A submission made by a prospective supplier in response to an invitation to tender. It makes an offer for the supply of goods or services.

Take Off

The term 'taking off' refers to identifying elements of construction works that can be measured and priced.

Tendering:

Tendering is when contractors and suppliers submit their prices and proposals for a construction project based on the BoQ and specifications. A competitive bidding process allows clients to receive multiple bids and select the most suitable contractor for their project. The tendering process ensures transparency, promotes fair competition, and helps clients secure the best value for their investment.

Unit Rate

"The cost per unit of measurement, like per square metre of floor area or cost per kilogram of steel reinforcement"

Unit Price Analysis:

Unit price analysis estimates project costs by analysing the cost per unit of a specific item or activity within the Bill of Quantities. It involves breaking down the BoQ into individual units and assessing the cost associated with each unit. This method allows for detailed cost estimation, accurate budgeting, and effective cost control.

Value Analysis:

Value analysis is a systematic approach to evaluating project components and their associated costs to identify opportunities for cost reduction without sacrificing quality. It aims to optimise the value of a construction project by assessing various design, material, and process alternatives. By taking a holistic view and considering factors such as functionality, aesthetics, and long-term efficiency, value analysis helps achieve the best possible outcome within the given cost parameters.

Value Engineering

is an essential practice in project management that aims to optimise the value of a project. It involves a systematic process of analysing various options to reduce costs without compromising quality or performance. By carefully evaluating every aspect of a project, value engineering helps identify potential areas where costs can be minimised while maintaining the desired level of quality. This approach goes beyond simply cutting expenses; it focuses on enhancing the overall value delivered by the project.

Variance

"Used to describe a difference. For example, the difference in planned vs actual cost or planned vs actual completion."

VOWD

Value of work to date: a project management technique for measuring and estimating the project cost at a point in time

Variation

An alteration to the scope of work initially specified in the contract, whether through an addition, omission, or substitution to the works or a change to how the works are to be carried out.

Vesting Certificate

An agreement for construction goods, plant or materials, in letter form, used to confirm that ownership of the goods, plant or materials will transfer from one party to another on payment

Work in Progress (WIP)

"Refers to cost or value of incomplete works in the construction process."

REFERENCE

Alaloul, W. S., Musarat, M. A., Rabbani, M. B. A., Iqbal, Q., Maqsoom, A., & Farooq, W. (2021). Construction sector contribution to economic stability: Malaysian GDP distribution. *Sustainability*, *13*(9), 5012.

Bharany, S., Sharma, S., Khalaf, O. I., Abdulsahib, G. M., Al Humaimeedy, A. S., Aldhyani, T. H., ... & Alkahtani, H. (2022). A systematic survey on energy-efficient techniques in sustainable cloud computing. *Sustainability*, *14*(10), 6256.

Boonsakul, P., Buddhawong, S., & Wangyao, K. (2022). Optimisation of multi-frequency electromagnetic surveying for investigating waste characteristics in an open dumpsite. *Journal of the Air & Waste Management Association*, *72*(11), 1290-1306.

Brinker, R. C., & Minnick, R. (2012). *The surveying handbook*. Springer Science & Business Media.

Costantino, D., Vozza, G., Pepe, M., & Alfio, V. S. (2022). Smartphone lidar technologies for surveying and reality modelling in urban scenarios:

evaluation methods, performance and challenges. *Applied System Innovation, 5*(4), 63.

Delgado, T., Bhark, S. J., & Donahue, J. (2021). Pandemic Teaching: Creating and teaching cell biology labs online during COVID-19. *Biochemistry and Molecular Biology Education, 49*(1), 32-37.

Dodanwala, T. C., Shrestha, P., & Santoso, D. S. (2021). Role conflict related job stress among construction professionals: The moderating role of age and organisation tenure. *Construction Economics and Building, 21*(4), 21-37.

Domone, P., & Illston, J. (2018). *Construction materials: their nature and behaviour.* CRC press.

Goodrum, P. M., Zhai, D., & Yasin, M. F. (2009). Relationship between changes in material technology and construction productivity. *Journal of construction engineering and management, 135*(4), 278-287.

Hao, M., Zhao, W., Qin, L., Mao, P., Qiu, X., Xu, L., ... & Qiu, G. (2021). A methodology to determine the optimal quadrat size for desert vegetation surveying based on unmanned aerial vehicle (UAV) RGB photography. *International Journal of Remote Sensing, 42*(1), 84-105.

Isakhodjayev, K., Mukhtarov, F., Kodirov, D., & Toshpulatov, I. (2021, December). Development of a laboratory nozzle chamber installation for the humidification of buildings. In *IOP Conference Series: Earth and Environmental Science* (Vol. 939, No. 1, p. 012025). IOP Publishing.

Khairadeen Ali, A., Lee, O. J., Lee, D., & Park, C. (2021). Remote indoor construction progress monitoring using extended reality. *Sustainability, 13*(4), 2290.

Linker, F. (1982). *PEP surveying procedures and equipment* (No. SLAC-TN-82-1). SLAC National Accelerator Lab., Menlo Park, CA (United States).

Lithgow, T., Stubenrauch, C. J., & Stumpf, M. P. (2023). Surveying membrane landscapes: a new look at the bacterial cell surface. *Nature Reviews Microbiology, 21*(8), 502-518.

Lopez, S. R., Bruun, G. R., Mader, M. J., & Reardon, R. F. (2021). The pandemic pivot: The impact of COVID-19 on mathematics and statistics post-secondary educators. *International Journal for*

Cross-Disciplinary Subjects in Education, 12(1), 4369-4378.

Matthews, R. A., Pineault, L., & Hong, Y. H. (2022). Normalizing the use of single-item measures: Validation of the single-item compendium for organizational psychology. *Journal of Business and Psychology, 37*(4), 639-673.

Myers, D. (2022). *Construction economics: A new approach.* Routledge.

Ofori, G. (2022). Introduction to the Research Companion to Construction Economics. In *Research Companion to Construction Economics* (pp. 1-17). Edward Elgar Publishing.

Provis, J. L., Duxson, P., & van Deventer, J. S. (2010). The role of particle technology in developing sustainable construction materials. *Advanced Powder Technology, 21*(1), 2-7.

Sly, P. G. (1981). Equipment and techniques for offshore survey and site investigations. *Canadian Geotechnical Journal, 18*(2), 230-249.

Tatum, C. B. (1988). Classification system for construction technology. *Journal of construction engineering and management, 114*(3), 344-363.

Yucesan, Y. A., Dourado, A., & Viana, F. A. (2021). A survey of modeling for prognosis and health management of industrial equipment. *Advanced Engineering Informatics, 50*, 101404.

Author Qualifications and Honours

D.D, Doctor of Divinity

Certificate in Bible studies

Theology

Laws

(LLM)Master of Law.

Postgraduate Laws

legal research,

Business, CSR Corporate social responsibility
and human rights law."

Institutional development and management,

International Law.

BA (Hons), Laws

Law: includes Criminal, Tort, damages,
Contract, Property, Equity and Trust, European Law,
Public, Constitutional, Judicial Review, and Agency.

Advance Dip. Business Law, Level 4:
Employment, Agency, Damages, Tort, Contract,
employment tribunal, etc.

Dip. Criminology

Accounting

BA (Hons)op.

Financial Accountant

Management Accountant

Cert. Acct; (Certified accountant)

(PCA)Professional Certificate in Financial and Management Accountant

Dip. Book-keeping, Level 3

Nursing

Nursing: RMN Registered Mental Nurse)

GN (General Trained Nurse)

Lecturer qualifications

DD Doctor of Divinity

LLM Master of Laws

BA (Hons)

BSc Hons o/g)psychology with counselling

Cert. in Education (Lecturer)

Business Certificate in Advanced Management

Cert. Business Enterprise

Advanced Food Hygiene

Intermediate Health and Safety

Dip. Safety Management

International Entrepreneur for over 25 years

(NVQ); Internal Verifier, (V1)

Trainer and Assessor A1 (NVQ)

Computers

Diploma: Cisco Level 2 Technician (build, repair, networking)

Microsoft Specialist

Dip. Claire Plus (in all software)

New Clait Dip. Level 2

ECDL Level 2

Scrip writer

Diploma in scrip writing.

TV, radio, stage, and film

Diploma in writing.

Autobiography

Biography

Family History

Certificate in Poetry

Psychology and Counselling

BSc(Hons o/g) psychology with counselling

Diploma in Counselling and Psychology

Cert. in Counselling and Psychology

Certificate in Social Science

Photography

Cert. (PGFP).

Portrait, Glamour and Figure

Plumbing

Level 3 City and Guild

Hypnotherapy

Dip. Hypnotherapy

Other Books by the Author James Safo

147 books.

Academic, Faith and non-faith books

Faith books- in 5 different languages :

Arabic, Chinese, English, French, Spanish

ALL FAITHS

Theology

Love All Faiths

Faith Unity

Religion And Law: religion influences National and international laws.

CHRISTIANITY

BIBLE New Testament; 1,111 QUESTIONS AND ANSWERS: Plus, synopsis and Test yourself

Bible Old Testament 1,064 Questions and Answers and Synopsis

Jesus Christ is Coming Soon

God Loves Christianity

God's/Allah's Messengers

Islam v. Christianity

Jesus Christ is Coming soon

Psychology of religion, politics and marriage

Faith unity.

Faith unity simplified version.

Islamisme versus Christianism.

Love all faith.

ISLAM (In English)

QUR'AN; 1,044 Questions & Answers.

Allah Loves Islam

Islam v. Christianity

BUDDHISM (In English)

God Enlighten Buddhism

HINDUISM (In English)

Parama Nandra Loves Hindus

FREEMASON (In English)

Quantity Survey   J. Safo

In Search of Wisdom in Freemasonry

FRENCH BOOKS (Religious)

Allah Aimel'islam (Allah loves Islam)

Aimetouteslesfois (Love All Faiths)

Islamism. V. christianisme (Islam v Christianity)

Dieu Aime Le Christianisme (God loves
Christianity)

Les Messagers De Dieu/ Allah (God/Allah
Messengers)

A LA Recherche De La Sagesse Dans La
Franc -Maconnerie (In Search of wisdom)

SPANISH BOOKS (Religious)

En Busca De Le Sabiduria Masoneri (In
search of wisdom in freemasonry)

Ametodas las creencias (Love All Faith)

Mensajeros de Dios (God Messengers)

4Dios Ama El Christianismo (God Loves Christianity)

Islamities v Cristianismo (Islam V Christianity)

Allah am el Islam (Allah Loves Islam)

ARABIC (Religious)

الله يحب الاسلام. . (Allah loves Islam)

حب جميع الأديان . Love All Faith

CHINESE BOOKS (Religious)

Books in Chinese

伊斯蘭教訴基督教 (Islam vs. Christianity)

上帝爱伊斯兰教 (Allah Loves Islam) -
Traditional Chinese Edition

Quantity Survey   J. Safo

NON-FAITH BOOKS- IN ENGLISH
LANGUAGES

LAW:

Global Injustice

The Journey to Law Graduation

THE JOURNEY TO MASTER OF LAWS

International Laws plus 30 dissertation

Laws - United Kingdom +30 dissertation

The Law (Over 1,160 Questions and Answers)

Business Law volume 1; over 800 Q&A
(contract, employment, types of Human Rights

Business Law Volume 2 over 600 Q&A (Tort,
CSR, Equity, Trust

Criminology: (Over 1,300 Questions and
Answers)

Religion And Law

POEMS

102 Poems on North America

70 Poems on South American Countries and Cities

80 POEMS ON THE ARCTIC AND ANTARCTICA

102 POEMS ON AUSTRALIA, OCEANIA, NEW ZEALAND

101 Poems on Asia countries and cities

118 USA POEMS: 50 States, Cities and Maps

114 Poems on 54 African countries

Over 200 Love Poems plus over 100 love icebreakers

Over 100 Poems on Faith & Victory

107 Poems on Discrimination, Racism & Suffering

Jesus Christ, Prophet, Arch Angels, Saint (Over 150 poems and Biography

The One - Over 130 Poems "DCF"

105 Poems on 54 European Countries & Cities

BUSINESS

Developing and Managing Institutions and Organisations Volume 1

Developing and Managing Institutions and Organisations Volume 2

Set up and manage a business

How to set up a care home and care agency

How to manage a care home and care agency

Care Home: Staff trainin

COMPUTER

Computing for beginners+310 questions and answers.

How to Build and Upgrade a Computer and Network

The Path of Information to the Computer Screen

Computer Programming, Coding & Science Dissertation

ACCOUNT

Financial Accounting (Over 1,241 Questions and Answers)

Management Accounting (1015 Questions & Answers Plus 100 Self-Assessment Questions)

Quantity Survey   J. Safo

PSYCHOLOGY

Journey to Psychology Graduation Volume 1

Journey to Psychology Graduation Volume 2

Psychology of Religion, Politics & Marriage

COUNSELLING

Journey to Counselling Graduation Volume 1

Journey to Counselling Graduation Volume 2

Counselling; Journey to Graduation Volume 3

Mood Disorder & Therapy

HISTORY

History; Journey to Graduation: 38 Essays

ENEMIES Within the earth

Slavery And Suffering

Slavery to Mastership

GEOGRAPHY

Quantity Survey   J. Safo

Geography: The Road to Graduation: 30 Essays

Medical/Nursing/ Health & Social

Drugs for Diseases: 1,007 Questions and Answers

Health and Social Care

Journey to Nursing Graduation: 51 Essays

Mental and Physical Diseases - Plus Nursing and 53 Dissertations

Health and Social Care - Plus 50 Dissertations

SOCIAL SCIENCE

Understanding Sociology Science - Plus 56 Dissertations

WOMEN

Women are superior to men

Sweet and Sour Women (plus over 500 love letters from women)

MANAGEMENT

Project management

RESEARCH

Research

Midwifery

A modern approach to Agriculture Introduction level,

Added Value to Agriculture advanced level.

The Roadmap to sustainable agriculture in rural development.

The beekeeper's blueprint: growing your apiary from the ground up

A modern approach to Agriculture Introduction level,

Added Value to Agriculture advanced level.

The Roadmap to sustainable agriculture in rural development.

Building and Construction

Quantity Surveying

Addictive Manufacturing

Drug abuse and Substance Rehabilitation

Index

www.ingramcontent.com/pod-product-compliance
Lightning Source LLC
Chambersburg PA
CBHW061138220326
41599CB00025B/4285